W0042310

Third Order Linear Differential Equations

Mathematics and Its Applications (*East European Series*)

Managing Editor:

M. HAZEWINKEL
Centre for Mathematics and Computer Science, Amsterdam, The Netherlands

Editorial Board:

A. BIALYNICKI-BIRULA, *Institute of Mathematics PKIN, Warsaw, Poland*
H. KURKE, *Humboldt University, Berlin, D.D.R.*
J. KURZWEIL, *Mathematics Institute, Czechoslovak Academy of Sciences, Prague, Czechoslovakia*
L. LEINDLER, *Bolyai Institute, Szeged, Hungary*
L. LOVÁSZ, *Eötvös Lóránd University, Budapest, Hungary*
D. S. MITRINOVIĆ, *University of Belgrade, Yugoslavia*
S. ROLEWICZ, *Polish Academy of Sciences, Warsaw, Poland*
BL. H. SENDOV, *Bulgarian Academy of Sciences, Sofia, Bulgaria*
I. T. TODOROV, *Bulgarian Academy of Sciences, Sofia, Bulgaria*
H. TRIEBEL, *University of Jena, D.D.R.*

Michal Greguš

Faculty of Mathematics and Physics, Comenius University, Bratislava, Czechoslovakia

Third Order Linear Differential Equations

Springer-Science+Business Media, B.V.
1987

Library of Congress Cataloging-in-Publication Data

Greguš Michal.

Third order linear differential equations.

(Mathematics and its applications. East European series).
Translation of: Lineárna diferenciálna rovnica tretieho rádu.
Bibliography: p.
Includes index.
1. Differential equations, Linear. I. Title.
II. Series: Mathematics and its applications (D. Reidel Publishing Company). East European series.

QA372.G73413 1986 515.3'54 86-3198
ISBN 978-94-010-8163-4 ISBN 978-94-009-3715-4 (eBook)
DOI 10.1007/ 978-94-009-3715-4

Scientific Editor
Prof. RNDr. Miloš Ráb, DrSc.

All rights reserved.
© 1987 by Michal Greguš
Originally published by D. Reidel Publishing Company in 1987
Softcover reprint of the hardcover 1st edition 1987
Translation © by J. Dravecký
No part of the material protected by this copyright notice may be reproduced or utilized in any form or by any means, electronic or mechanical, including photocopying, recording or by any information storage and retrieval system without written permission from the copyright owner.

SERIES EDITOR'S PREFACE

Approach your problems from the right end and begin with the answers. Then one day, perhaps you will find the final question.

'The Hermit Clad in Crane Feathers' in R. van Gulik's *The Chinese Maze Murders*.

It isn't that they can't see the solution. It is that they can't see the problem.

G. K. Chesterton. *The Scandal of Father Brown* 'The Point of a Pin'.

Growing specialization and diversification have brought a host of monographs and textbooks on increasingly specialized topics. However, the "tree" of knowledge of mathematics and related fields does not grow only by putting forth new branches. It also happens, quite often in fact, that branches which were thought to be completely disparate are suddenly seen to be related.

Further, the kind and level of sophistication of mathematics applied in various sciences has changed drastically in recent years: measure theory is used (non-trivially) in regional and theoretical economics; algebraic geometry interacts with physics; the Minkowsky lemma, coding theory and the structure of water meet one another in packing and covering theory; quantum fields, crystal defects and mathematical programming profit from homotopy theory; Lie algebras are relevant to filtering; and prediction and electrical engineering can use Stein spaces. And in addition to this there are such new emerging subdisciplines as "experimental mathematics", "CFD", "completely integrable systems", "chaos, synergetics and large-scale order", which are almost impossible to fit into the existing classification schemes. They draw upon widely different sections of mathematics. This programme, Mathematics and Its Applications, is devoted to new emerging (su)disciplines and to such (new) interrelations as exempli gratia:

— a central concept which plays an important role in several different mathematical and/or scientific specialized areas;
— new applications of the results and ideas from one area of scientific endeavour into another;
— influences which the results, problems and concepts of one field of enquiry have and have had on the development of another.

The Mathematics and Its Applications programme tries to make available a careful selection of books which fit the philosophy outlined above. With such books, which are stimulating rather than definitive, intriguing rather than encyclopaedic, we hope to contribute something towards better communication among the practitioners in diversified fields.

Because of the wealth of scholarly research being undertaken in the Soviet Union, Eastern Europe, and Japan, it was decided to devote special attention to the work emanating from these particular regions. Thus it was decided to start three regional series under the umbrella of the main MIA programme.

As a general (and therefore not very profound) rule, a field of research needs two things to flourish: general, deep and far-reaching concepts and detailed, thorough, exhaustive studies of particular classes of examples or problems. It is often perfectly amazing (for a relative outsider) to see how many powerful concepts go back finally to one good, well-studied, particular example. One famous illustration of this is Fermat's "last theorem", the conjecture which states that $x^n + y^n = 2^n$ for $n \geqslant 3$ has no non-trivial integer solutions, which almost alone gave birth to the field of algebraic number theory. This example in itself has nothing to do with the subject matter of this book. Another case in point is the topic of second order ordinary differential equations. Now, third order ordinary differential equations, the subject of this book, behave very differently from second order ones. New, very different phenomena appear (bands of solutions) and they will likely be the basis and starting point for several new developments. This book is basically a survey of what is known about third order differential equations. As such it is a unique supplement to the existing literature on ordinary differential equations which, as a rule, only treats second order ones and certain special higher even order differential equations.

The unreasonable effectiveness of mathematics in science ...

Eugene Wigner

Well, if you know of a better 'ole, go to it.

Bruce Bairnsfather

What is now proved was once only imagined.

William Blake

As long as algebra and geometry proceeded along separate paths, their advance was slow and their applications limited.

But when these sciences joined company they drew from each other fresh vitality and thenceforward marched on at a rapid pace towards perfection.

Joseph Louis Lagrange

Bussum, March 1986 Michiel Hazewinkel

Contents

Preface

This book contains the theory of third order linear homogeneous differential equations in the so-called normal form and a survey of the most important results in the theory of third order linear homogeneous differential equations with continuous coefficients.

The majority of books dealing with the theory of ordinary differential equations, or their practical application to technology and physics, contain at most the results of the second order linear differential equation theory and possibly some results concerning the theory of some special linear differential equations of higher, even orders. Monographs (Swanson [139], Mckelvey [96]), which include chapters on oscillation properties of third order differential equations, are exceptional.

No exacting mathematical apparatus is necessary for the study of linear differential equation theory, and therefore the present book on the theory of third order linear homogeneous differential equations can be successfully read by those who have completed their (professional or teachers) study of mathematics and/or physics at Universities, even by advanced students or graduates of Technical Colleges who are acquainted with the principles of ordinary differential equation theory and of mathematical analysis.

The first paper, which can be regarded as classical, on third order differential equations appeared in 1910 (Birkhoff [12]) in which methods of projective geometry were applied to such equations by G. D. Birkhoff. In the subsequent period, although the theory of linear differential equations of the second and also of the fourth order was intensively developed, third order linear differential equations were rarely mentioned in literature [2, 19, 24, 25].

It was as late as 1948 that Sansone [125] summarized the results which had already been published on the third order differential equations and substantially enriched this theory by new results. He formulated many so far unsolved problems, especially in the field of boundary value problems and comparison theorems, thus giving a strong impetus towards further development of the theory. Professor Borůvka in 1950 attracted many young mathematicians from Brno and Bratislava to his seminar in Brno on the theory of dispersions and transformations of the second order linear differential equation. There he also pointed out to a number of unsolved problems in third order differential equation theory. These problems were then intensively studied and discussed both in that mentioned seminar and in a seminar at the Comenius University in Bratislava besides, of course, other problems in differential equation theory.

In addition to the results achieved by the participants in these two seminars, further key results are worth noting, namely those published in Azbelev [4], Azbelev and Tsalyuk [5], Barrett [7, 8], Hanan [61], Jones [69—74], Kondratev [79, 80], Lazer [86], Vilari [142, 143] and in other papers.

The original edition of this book, in Slovak, has three chapters. The first chapter contains the theory of the so-called normal form of the third order differential equation, especially oscillatory and asymptotic properties of solutions, as well as the theory of multi-point boundary value problems. In the second chapter, a study is made of the theory of third order linear homogeneous differential equations with continuous coefficients. Besides introductory sections written in some detail, it includes an up-to-date survey of the most important results of the theory. The short Chapter 3 contains remarks on the study of certain special types of third order differential equations and a note about the theory of transformations of such equations. In the present English edition, Chapter 4 on applications of the third order linear differential equation theory has been added.

A unifying notion of the whole theory is that of a band of solutions and the properties of bands as two-parameter systems of solutions satisfying a certain second order differential equation.

Finally, I wish to thank my teacher, Academician of the Czechoslovak Academy of Sciences O. Borůvka, for bringing the topic to my attention and valuable advice while studying it. I am also obliged to Assoc. Prof. RNDr. M. Gera, Assoc. Prof. RNDr. F. Neuman and Prof. RNDr. M. Ráb for kindly reading and thoroughly checking the manuscript.

Author

Chapter I

Third Order Linear Homogeneous Differential Equations in Normal Form

§1. FUNDAMENTAL PROPERTIES OF SOLUTIONS OF THE THIRD ORDER LINEAR HOMOGENEOUS DIFFERENTIAL EQUATION

1. The Normal Form of a Third Order Linear Homogeneous Differential Equation

Let a third order linear differential equation be given in the form

$$u''' + 3p_1u'' + 3p_2u' + p_3u = 0 ,$$

where $p_1 = p_1(x)$, $p_2 = p_2(x)$, $p_3 = p_3(x)$ are functions defined on an interval (a, b), $-\infty \leqq a$, $b \leqq \infty$, and let p_1'', p_2', p_3 be continuous functions of $x \in (a, b)$, $'$ denoting the derivative of a function with respect to the independent variable. Let $x_0 \in (a, b)$. By the transformation $u = y \exp\left(-\int_{x_0}^{x} p_1(t)\, dt\right)$, the above differential equation takes the form

$$y''' + 3P_2y' + P_3y = 0 ,$$

where $P_3 = p_3 - 3p_1p_2 + 2p_1^3 - p_1''$, $P_2 = p_2 - p_1^2 - p_1'$. By writing

$$A = \frac{3}{2}P_2 , \quad b = P_3 - \frac{3}{2}P_2' ,$$

we obtain the following differential equation for y,

$$y''' + 2Ay' + (A' + b)y = 0 . \qquad \text{(a)}$$

The functions $A = A(x)$, $A' = A'(x)$, $b = b(x)$ are evidently continuous for $x \in (a, b)$. Throughout Chapter I the coefficients A', b will be assumed to be continuous whenever the equation (a) is under consideration and this assumption will not be repeatedly stated.

The form (a) of a differential equation is called the normal form (Birkhoff [12]) and the function $b = b(x)$ is referred to as the Laguerre invariant (Laguerre [82]).

By a solution of a third order differential equation we mean a function $y = y(x)$ having a continuous third derivative and satisfying the given differential equation in the relevant interval.

2. *Adjoint and Self-adjoint Third Order Linear Differential Equations*

The differential equation adjoint to (a) (Sansone [124]) has the form

$$z''' + 2Az' + (A' - b)z = 0 . \tag{b}$$

If y_1, y_2 are any two linearly independent solutions of the differential equation (a), then $z = y_1 y_2' - y_1' y_2$ is a solution of the differential equation (b). This assertion is verified as follows.

Obviously

$$\begin{aligned} y_1''' + 2Ay_1' + (A' + b)y_1 &= 0 , \\ y_2''' + 2Ay_2' + (A' + b)y_2 &= 0 . \end{aligned} \tag{1.1}$$

On multiplying the first equation of (1.1) by y_2 and the second equation by y_1, and subtracting, we get

$$(y_2 y_1''' - y_1 y_2''') + 2A(y_2 y_1' - y_2' y_1) = 0, \text{ i.e.}$$

$$\begin{vmatrix} y_1 , & y_2 \\ y_1''', & y_2''' \end{vmatrix} = -2Az .$$

On differentiating twice with respect to z we obtain

$$z'' = \begin{vmatrix} y_1 , & y_2 \\ y_1''', & y_2''' \end{vmatrix} + \begin{vmatrix} y_1', & y_2' \\ y_1'', & y_2'' \end{vmatrix} ,$$

whence

$$z'' + 2Az = \begin{vmatrix} y_1', & y_2' \\ y_1'', & y_2'' \end{vmatrix} .$$

Differentiating the last equality term by term we get

$$z''' + 2A'z + 2Az' = \begin{vmatrix} y'_1 , & y'_2 \\ y'''_1 , & y'''_2 \end{vmatrix} . \tag{1.2}$$

If we now multiply the first equation of (1.1) by y_2' and the second one by y_1', we obtain, after subtraction,

$$\begin{vmatrix} y_1' , & y_2' \\ y_1''' , & y_2''' \end{vmatrix} = (A' + b)z .$$

Substituting from the last equality into (1.2) we verify the assertion.

If y_1, y_2, y_3 form a fundamental system of solutions of the differential equation (a) whose wronskian $W(y_1, y_2, y_3) = 1$, then the functions

$$z_1 = \begin{vmatrix} y_2, & y_3 \\ y_2', & y_3' \end{vmatrix} , \quad z_2 = \begin{vmatrix} y_3, & y_1 \\ y_3', & y_1' \end{vmatrix} , \quad z_3 = \begin{vmatrix} y_1, & y_2 \\ y_1', & y_2' \end{vmatrix}$$

form a fundamental set of solutions of the differential equation (b) (Sansone [124]). The converse proposition is evidently true also. Using a suitable fundamental system of the differential equation (b), we can construct a fundamental system of solutions of (a).

If $b(x) \equiv 0$ for $x \in (a, b)$, the differential equation (a) or (b) becomes a self-adjoint third order differential equation of the form

$$y''' + 2Ay' + A'y = 0 . \tag{1.3}$$

Let u be a solution of the second order differential equation

$$u'' + \frac{1}{2} Au = 0 , \tag{1.4}$$

then the function $y = u^2$ is a solution of the differential equation (1.3). This proposition is easily proved by substitution in (1.3). In fact, $y' = 2uu'$, $y'' = 2u'^2 + 2uu''$, $y''' = 6u'u'' + 2uu'''$. From this and the equation (1.3) it follows that

$$2uu''' + 6u'u'' + 4Auu' + A'u^2 =$$

$$2u\left(u''' + \frac{1}{2} Au' + \frac{1}{2} A'u\right) + 6u'\left(\frac{1}{2} Au + u''\right) = 0 ,$$

since

$$u''' + \frac{1}{2} Au' + \frac{1}{2} A'u = \left(u'' + \frac{1}{2} Au\right)' = 0 .$$

It can similarly be proved that if u and v are linearly independent solutions of the differential equation (1.4), then the function $y = uv$ satisfies the equation (1.3).

The functions u^2, uv, v^2 form a fundamental set of solutions of the differential equation (1.3) (Birkhoff [12] or Sansone [124]).

3. Fundamental Properties of Solutions

From general theorems on existence and uniqueness of solutions to differential systems it follows that corresponding to any quadruple (x_0, y_0, y_0', y_0'') of numbers $x_0 \in (a, b)$, $-\infty < y_0 < \infty$, $-\infty < y_0' < \infty$, $-\infty < y_0'' < \infty$ there is exactly one solution y of the differential equation (a) having the property $y(x_0) = y_0$, $y'(x_0) = y_0'$, $y''(x_0) = y_0''$ and defined on the whole interval (a, b) (Sansone [124]).

We easily prove the following theorem.

THEOREM 1.1. The solutions of the differential equation (a) with the property $y(x_0) = y'(x_0) = 0$, $y''(x_0) \neq 0$, $x_0 \in (a, b)$ are linearly dependent.

PROOF. Let y_1, y_2, y_3 be a fundamental system of solutions of the differential equation (a). Then the function

$$Y(x) = \begin{vmatrix} y_1(x_0), & y_2(x_0), & y_3(x_0) \\ y_1'(x_0), & y_2'(x_0), & y_3'(x_0) \\ y_1(x), & y_2(x), & y_3(x) \end{vmatrix}$$

is evidently a solution of the differential equation (a) and satisfies $Y(x_0) = Y'(x_0) = 0$, $Y''(x_0) \neq 0$. let $y(x)$ be an arbitrary solution of the differential equation (a) with the property $y(x_0) = y'(x_0) = 0$, $y''(x_0) = \beta \neq 0$. Then there exist constants c_1, c_2, c_3 such that $y = c_1 y_1 + c_2 y_2 + c_3 y_3$. If we calculate these constants, taking into account the initial conditions, we get $y(x) = (\beta / Y''(x_0))\, Y(x)$. Thus the theorem is proved. □

It can be proved similarly that the solutions of the differential equation (a) with the property

$$y(x_0)=y''(x_0)=0\ ,\quad y'(x_0)\neq 0$$

or

$$y'(x_0)=y''(x_0)=0,\quad y(x_0)\neq 0$$

are also linearly dependent.

Let the functions y_1, y_2, y_3 form a fundamental set of solutions of the differential equation (a) satisfying

$$y_1(x_0)=y_1'(x_0)=0,\quad y_1''(x_0)\neq 0,\quad y_2(x_0)=y_2''(x_0)=0\ ,$$

$$y_2'(x_0)\neq 0,\quad y_3'(x_0)=y_3''(x_0)=0,\quad y_3(x_0)\neq 0\ ,$$
$$x_0\in(a,\,b)\ . \tag{1.5}$$

Then we have the following theorem.

THEOREM 1.2. Every solution y of the differential equation (a) with the property $y(x_0)=0$, $x_0\in(a,\,b)$ can be expressed in the form $y=c_1y_1+c_2y_2$, where c_1, c_2 are suitable constants.

PROOF. Let $y(x)$ be a solution of the differential equation (a) with $y(x_0)=0$, $y'(x_0)=\alpha$, $y''(x_0)=\beta$, where at least one of the numbers α, β is non-zero. We show that c_1 and c_2 in $y=c_1y_1+c_2y_2$ can be chosen so that y satisfies the initial conditions. The first condition $y(x_0)=0$ is satisfied identically for every c_1, c_2. From the second condition $y'(x_0)=c_2y_2'(x_0)=\alpha$ it follows that $c_2=\alpha/y_2'(x_0)$, and the third condition $y'(x_0)=c_1y_1(x_0)=\beta$ yields $c_1=\beta/y_1''(x_0)$, so the theorem is proved. □

We can prove similarly that every solution y of the differential equation (a) satisfying $y'(x_0)=0$ or $y''(x_0)=0$, $x_0\in(a,\,b)$, may be written as $y=c_1y_1+c_2y_3$ or $y=c_1y_2+c_2y_3$, where c_1, c_2 are suitable constants.

4. *Relationship between Solutions of the Differential Equations (a) and (b)*

Besides the relations between solutions of differential equations adjoint to each other, which were given in Section 2, there are some relations between the zeros of solutions of mutually adjoint differential equations.

THEOREM 1.3. A necessary and sufficient condition for a solution y of the differential equation (a) with $y(x_1)=y'(x_1)=0$, $y''(x_1)\neq 0$, $x_1\in(a,b)$ to have another zero at $x_2\neq x_1\in(a,b)$ is that x_1 be a zero of a solution z of the differential equation (b) with $z(x_2)=z'(x_2)=0$, $z''(x_2)\neq 0$.

PROOF. Necessity. Let y be a solution of (a) with $y(x_1)=y'(x_1)=0$, $y''(x_1)\neq 0$, $y(x_2)=0$, $x_1\neq x_2\in(a,b)$. By Theorem 1.2, y may be expressed in the form $y=c_1y_1+c_2y_2$, where y_1, y_2 satisfy (1.5) at $x_0=x_2$. The constants c_1, c_2 are to be chosen to meet the following equations

$$c_1y_1(x_1)+c_2y_2(x_1)=0\,,$$
$$c_1y_1'(x_1)+c_2y_2'(x_1)=0$$

and at least one of these constants must be non-zero. This requires that the determinant of the system must vanish, that is to say, $z(x_1)=y_1(x_1)y_2'(x_1)-y_1'(x_1)y_2(x_1)=0$; however, $z=y_1y_2'-y_1'y_2$ is a solution of (b) with a double zero at x_2.

Sufficiency. Let z be a solution of (b) with $z(x_2)=z'(x_2)=0$, $z''(x_2)\neq 0$, $z(x_1)=0$, $x_2\neq x_1\in(a,b)$ and let y be a solution of (a) satisfying $y(x_1)=y_1'(x_1)=0$, $y''(x_1)\neq 0$. We have to show that $y(x_2)=0$. Let y_1, y_2, y_3 be a fundamental set of solutions of the differential equation (a) satisfying (1.5) at $x_0=x_2$. Then y may be written in the form $y=c_1y_1+c_2y_2+c_3y_3$, where c_1, c_2, c_3 are to be chosen so that

$$c_1y_1(x_1)+c_2y_2(x_1)+c_3y_3(x_1)=0\,,$$
$$c_1y_1'(x_1)+c_2y_2'(x_1)+c_3y_3'(x_1)=0\,,$$
$$c_1y_1''(x_1)+c_2y_2''(x_1)+c_3y_3''(x_1)=k\neq 0\,,$$

where k is a suitable constant. The above equations yield for c_3 the value

$$c_3=\frac{k}{W(x_1)}\,[y_1(x_1)y_2'(x_1)-y_1'(x_1)y_2(x_1)]=\frac{k}{W(x_1)}\,z(x_1)=0\,.$$

Here, $W(x_1)$ is the wronskian of the fundamental set of solutions y_1, y_2,

y_3. Hence it follows that $y = c_1y_1 + c_2y_2$ and therefore $y(x_2) = 0$, which proves the theorem. □

The following two theorems can be proved similarly.

THEOREM 1.4. A necessary and sufficient condition for a solution y of (a) with $y(x_1) = y''(x_1) = 0$, $y'(x_1) \neq 0$, $x_1 \in (a, b)$ to have another zero at $x_2 \neq x_1 \in (a, b)$ is that $z'(x_1) = 0$, where z is a solution of (b) with a double zero at x_2.

THEOREM 1.5. A necessary and sufficient condition for a solution y of (a) with $y'(x_1) = y''(x_1) = 0$, $y(x_1) \neq 0$, $x_1 \in (a, b)$ to have another zero at $x_2 \neq x_1 \in (a, b)$ is that x_1 be a zero of the function $z'' + 2Az$, where z is a solution of (b) having a double zero at x_2.

5. Integral Identities

The following integral identities (Sansone [125]) are true for solutions of the differential equation (a)

$$yy'' - \frac{1}{2}y'^2 + Ay^2 + \int_{x_0}^{x} by^2 \, dt = \text{const}, \tag{1.6}$$

$$y'' + 2Ay + \int_{x_0}^{x} (b - A')y \, dt = \text{const}, \tag{1.7}$$

where $x_0 \in (a, b)$ is a fixed number, $x \in (a, b)$ is variable.

The integral identity (1.6) is obtained by multiplying the differential equation (a) by y and integrating termwise from x_0 to x.

Similarly, we obtain the integral identity (1.7) by integrating the differential equation (a) term by term from x_0 to x.

The solutions of the differential equation (b) satisfy analogous integral identities

$$zz'' - \frac{1}{2}z'^2 + Az^2 - \int_{x_0}^{x} bz^2 \, dt = \text{const}, \tag{1.8}$$

and

$$z'' + 2Az - \int_{x_0}^{x} (A' + b)z \, dt = \text{const}. \tag{1.9}$$

The integral identity (1.6) implies the following simple result.

If $b(x) \geqq 0$ for $x \in (a, b)$ and $b(x) \not\equiv 0$ in any subinterval of (a, b),* then no solution y of the differential equation (a) with $y(x_1) = y'(x_1) = 0$, $y''(x_1) \neq 0$, $x_0 \in (a, b)$ has a zero to the left of x_0 in (a, b).

A similar property is possessed by every solution z of the differential equation (b) to the right of a double zero.

6. *Notion of a Band of Solutions of the First, Second and Third Kinds*

Let y_1, y_2, y_3 be a fundamental system of solutions of the differential equation (a) with the property (1.5).

A set of solutions of the differential equation (a), $y = c_1 y_1 + c_2 y_2$, with $y(x_1) = 0$ is called a *band of solutions of the first kind* of the differential equation (a) at x_0, or briefly a *band of the first kind* at x_0.

Analogously, a set of solutions $y = c_1 y_1 + c_2 y_3$ with $y'(x_1) = 0$ will be referred to as a *band of solutions of the second kind* of the differential equation (a) at x_0, or shortly a *band of the second kind* at x_0.

Finally, a set of solutions $y = c_1 y_2 + c_2 y_3$ with $y''(x_1) = 0$ is called a *band of solutions of the third kind* of the differential equation (a) at x_0, concisely, a *band of the third kind* at x_0.

THEOREM 1.6. Every band of solutions of the differential equation (a) satisfies the second order differential equation

$$wy'' - w'y' + (w'' + 2aw)y = 0\,, \tag{c}$$

where w is a solution of the differential equation (b), namely $w = w_1 = y_1 y_2' - y_1' y_2$ in case of a band of the first kind, $w = w_2 = y_1 y_3' - y_1' y_3$ in case of a band of the second kind, and $w = w_3 = y_2 y_3' - y_2' y_3$ in case of a band of the third kind.

Theorem 1.6 is proved without difficulty by eliminating the constants c_1, c_2 from appropriate equations, e.g., in case of a band of the first kind the equations to be used are

* We use the symbol $b(x) \not\equiv 0$ on an interval I to mean that $b(x)$ is not identically zero on I; that is to say, there exists at least one point of I for which $b(x) \neq 0$.

$$y = c_1 y_1 + c_2 y_2 ,$$
$$y' = c_1 y_1' + c_2 y_2' ,$$
$$y'' = c_1 y_1'' + c_2 y_2'' .$$

REMARK 1.1. The operator on the left-hand side of the differential equation (c) can also be obtained as a linear form from the Lagrange identity giving the relationship between differential equations adjoint to each other (Sansone [124]).

REMARK 1.2. If $w \neq 0$ in some interval $I \subset (a, b)$, then the zeros of solutions of the respective band in that interval (if they exist) separate each other.

It follows from the integral identity (1.8) that if $b(x) \geqq 0$ for $x \in (a, b)$ and $b(x) \not\equiv 0$ in any subinterval of (a, b), then the solution w_1 of differential equation (b) has no zero in (a, b) to the right of x_0, since $w_1(x_0) = w_1'(x_0) = 0$, $w_1''(x_0) \neq 0$.

THEOREM 1.7. Let $b(x) \geqq 0$ for $x \in (a, b)$ and let $b(x) \not\equiv 0$ in any subinterval of (a, b). Assume also that $A(x) \leqq 0$ for $x \in (a, b)$. Then $w_3(x) \neq 0$ for $x > x_0 \in (a, b)$. If $A(x) \leqq 0$ and $A'(x) + b(x) \geqq 0$ for $x \in (a, b)$, then $w_2(x) \neq 0$ for $x > x_0 \in (a, b)$.

PROOF. Suppose that $w_2(x_0) = w_2''(x_0) = 0$, $w_2'(x_0) \neq 0$ and $w_3(x_0) \neq 0$, $w_3'(x_0) = 0$, $w_3''(x_0) = -2A(x_0) w_3(x_0)$.

The integral identity (1.8) for w_3 takes the following form

$$w_3 w_3'' - \frac{1}{2} w_3'^2 + A w_3^2 - \int_{x_0}^{x} b w_3^2 \, dt =$$
$$= -2a(x_0) w_3^2(x_0) + A(x_0) w_3^2(x_0) \geqq 0 .$$

If we assume that $w_3(x_1) = 0$, $x_1 > x_0 \in (a, b)$, the above identity yields a contradiction, hence $w_3(x) \neq 0$ for $x > x_0$.

The solution w_2 evidently satisfies the identity (1.9) in the form

$$w_2'' + 2A w_2 - \int_{x_0}^{x} (A' + b) w_2 \, dt = 0 .$$

Clearly, we have

$$(w_2 w_2')' = w_2 w_2'' + w_2'^2 .$$

Substituting in the above equality for w_2'' from the preceding identity we get

$$(w_2 w_2') = -2Aw_2^2 + w_2 \int_{x_0}^{x} (A' + b) w_2 \, dt + w_2'^2 .$$

The expression on the right-hand side is non-negative for $x > x_0 \in (a, b)$. If we assume that $w_2(x_1) = 0$ for $x_1 > x_0 \in (a, b)$, a contradiction will arise after integrating from x_0 to x_1. □

REMARK 1.3. Using the above identities it can be proved that if $b(x) \geqq 0$ for $x \in (a, b)$ and $b(x) \not\equiv 0$ in any subinterval of (a, b), and moreover, if

a) $A'(x) + b(x) > 0$ for $x \in (a, b)$, then $w_i''(x) + 2Aw_i(x) \neq 0$, $i = 1, 2$, for $x > x_0$,

b) $A(x) \leqq 0$, $A'(x) + b(x) \geqq 0$ for $x \in (a, b)$, then $W_2'(x) \neq 0$ for $x > x_0$,

c) $A(x) \leqq 0$, $A'(x) + b(x) > 0$ for $x \in (a, b)$, then $w_3''(x) + 2A(x) w_3(x) \neq 0$ for $x > x_0$,

d) $A(x) \leqq 0$ for $x \in (a, b)$, then $w_1(x) \neq 0$ and $w_1'(x) \neq 0$ for $x > x_0$.

COROLLARY 1.1. Let $b(x) \geqq 0$ for $x \in (a, b)$ and let $b(x) \not\equiv 0$ in every subinterval of (a, b). Then the solutions of the differential equation (a) with $y(x_0) = y^{(i)}(x_1) = 0$, $x_0 < x_1 \in (a, b)$, for every $i = 0, 1, 2$, are linearly dependent. If we assume, moreover, that $A(x) \leqq 0$ for $x \in (a, b)$ [$A(x) \leqq 0$ and $A'(x) + b(x) \geqq 0$ for $x \in (a, b)$], then the solutions of the differential equation (a) with $y(x_0) = y^{(i)}(x_1) = 0$ [$y'(x_0) = y^{(i)}(x_1) = 0$], $x_0 < x_1 \in (a, b)$, $i = 0, 1, 2$, are also linearly dependent.

THEOREM 1.8. Let $b(x) \geqq 0$ for $x \in (a, b)$ and let $b(x) \not\equiv 0$ in any subinterval of (a, b). Let $x_1 > x_0 \in (a, b)$ be the first zero of y_1. Then every solution y in the band of the first kind at x_0 which is independent of y_1 has exactly one zero between x_0 and x_1.

PROOF. It is sufficient to show that the interval (x_0, x_1) contains at least one zero of a given solution y from the band of the first kind at x_0. Since the zeros to the right of x_0 of solutions in the band of the first kind separate each other, there is exactly one zero of y in the interval (x_0, x_1).

Suppose the contrary. Then

$$\left(\frac{y_1}{y}\right)' = \frac{y_1'y - y_1y'}{y^2}.$$

Integrating this equality from x_0 to x_1 we get

$$\frac{y_1(x_1)}{y(x_1)} - \lim_{x \to x_0^+} \frac{y_1(x)}{y(x)} = \lim_{x \to x_0^+} \int_x^{x_1} \frac{y_1'y \quad y_1y'}{y^2}\, dt.$$

The left-hand side of the last equality is zero. The integral on the right-hand side exists, because the integrand is continuous in (x_0, x_1) and has a limit at x_0. Since $y_1'y - y_1y' = kw_1 \neq 0$ in (x_0, x_1), the integral on the right-hand side is non-zero, which is a contradiction, so the theorem is proved. □

THEOREM 1.9. Let $b(x)$ satisfy the same hypotheses as in Theorem 1.8. Let $x_1 > x_0 \in (a, b)$ be the first zero of the solution y_1 of the differential equation (a) having a double zero at x_0. Let $x_0 < \bar{x} < x_1$ and let $\bar{y}_1$ be a solution of (a) having a double zero at $\bar{x}$, and let $\bar{x}_1 > \bar{x} \in (a, b)$ be its next zero. Then $\bar{x}_1 > x_1$.

PROOF. Suppose the contrary, i.e., let $x_0 < \bar{x} < \bar{x}_1 \leqq x_1$. Then a solution y_{x_0} belonging to the band at x_0 begins at x_0 and passes through $\bar{x}$, that is, it belongs to the band at $\bar{x}$ as well. According to Theorem 1.8, it has another zero between $\bar{x}$ and $\bar{x}_1$. This means that it has two zeros between x_0 and x_1, contradicting the assertion of Theorem 1.8. □

COROLLARY 1.2. Let $x_1 < x_2$ be two neighbouring zeros of a solution of the differential equation (a) and let $b(x)$ satisfy the same hypotheses as in Theorem 1.8. Then no solution $\bar{y}$ of the differential equation (a) with $\bar{y}(\bar{x}) = \bar{y}(x_2) = 0$, $x_1 < \bar{x} < x_2 \in (a, b)$, has another zero between x_1 and x_2.

COROLLARY 1.3. Let $b(x)$ satisfy the assumptions stated in Theorem 1.8. Then a necessary and sufficient condition for a solution y of the differential equation (a) with $y(x_1)=y(x_2)=0$, $x_1<x_2\in(a, b)$, not to have another zero between x_1 and x_2, is that

$$0<x_2-x_1\leqq x_2-x_\alpha ,$$

where x_α is the first zero to the left of x_2 of the solution w_1 of the differential equation (b) having a double zero at x_2 (assuming that x_α exists).

PROOF. Necessity. Let y be a solution of the differential equation (a) with $y(x_1)=y(x_2)=0$, $x_1<x_2\in(a, b)$, having no more zeros between x_1 and x_2. Further, let w_1 be a solution of the differential equation (b) satisfying $w_1(x_2)=w_1'(x_2)=0$, $w_1''(x_2)\neq 0$ and such that x_α is its first zero to the left of x_2. Suppose that the assertion of Corollary 1.3 is false and that $x_1<x_\alpha<x_2$. It follows then from Theorem 1.3 that there exists a solution of the differential equation (a) with a double zero at x_α and vanishing at x_2. In the band of the first kind at x_1 of the differential equation (a) there exists exactly one solution (up to linear dependence) having a zero at x_α and necessarily having another zero between x_α and x_2, which is a contradiction.

Sufficiency. Suppose that $0<x_2-x_1\leqq x_2-x_\alpha$, where x_1, x_2 are zeros of a solution y of (a) and x_α is a zero of the solution w_1 of (b) having a double zero at x_2. Let y have another zero between x_1 and x_2. It follows from Theorem 1.3 that there is a solution Y of the differential equation (a) having a double zero at x_α and vanishing at x_2. If Y has no other zero between x_α and x_2, it follows from Corollary 1.2 that y also has no zeros between x_1 and x_2. We show that Y has no zeros between x_α and x_2. The solution Y may be written in the form $Y=c_1y_1+c_2y_2$, where y_1 and y_2 are solutions of (a) satisfying (1.5) at $x_0=x_2$. The solution y_1 has no other zero to the left of x_2. Assume that $Y(\bar{x})=0$, $x_\alpha<\bar{x}<x_2$. Evidently,

$$\left(\frac{c_1y_1+c_2y_2}{y_1}\right)'=c_2\frac{kw_1}{y_1^2}, \quad k\neq 0 \text{ for } y_1(x)\neq 0 .$$

Integrating the last equality from x_α to $\bar{x}$ leads to a contradiction. Therefore, Y cannot have a zero at $\bar{x}$, and the theorem is proved. □

7. *Further Properties of Solutions of the Differential Equation (a) Implied by Properties of Bands*

From now on, by a band of solutions we shall mean a band of the first kind (Greguš [38, 44, 46]).

Throughout this section, we shall assume that the function b has the following property

$$b(x) \geqq 0 \text{ for all } x \in (a, b) \text{ while } b(x) \not\equiv 0 \tag{v}$$

in any subinterval.

THEOREM 1.10 Let $\alpha \in (a, b)$. Then a necessary and sufficient condition in order that every solution of the differential equation (a) having a zero should have infinitely many zeros in (α, b), is the existence of at least one solution of the differential equation (a) which has infinitely many zeros in (α, b).

PROOF. Necessity is obvious. Sufficiency follows from the properties of bands. Let $\bar{y}(x)$ be a solution of (a) having infinitely many zeros in (α, b). Let $y(x)$ be any solution of (a) vanishing at $x_1 \in (a, b)$ and let $\bar{x} > x_1 \in (a, b)$ be a zero of $\bar{y}$. Some solution in the band at x_1 vanishes at $\bar{x}$. This solution, together with $\bar{y}$, belongs to the band at $\bar{x}$, therefore it has infinitely many zeros in $(\bar{x}, b)$. Hence y must also have infinitely many zeros in (α, b), which proves the theorem. □

The following two theorems can be proved similarly.

THEOREM 1.11. A necessary and sufficient condition in order that every solution of the differential equation (a) having a zero should have infinitely many zeros in (α, b), $a < \alpha < b$, is that every solution of (a) with a double zero in (α, b) has at least one other zero.

To prove Theorem 1.11 it is necessary to use Theorem 1.8.

THEOREM 1.12. If a solution $\bar{y}$ of the differential equation (a) with a double zero at $x_0 \in (a, b)$ has no other zero, then no solution of (a) having a double zero vanishes again on the right of x_0.

REMARK 1.4. Theorems 1.10 and 1.11 may be formulated for the differential equation (b) as well. We must bear in mind, however, that $-b(x) \leqq 0$ for $x \in (a, b)$.

LEMMA 1.1. Let w be any solution of the differential equation (b). Then there exist two solutions y_1, y_2 of the differential equation (a) such that $w = y_1 y_2' - y_1' y_2$ for all $x \in (a, b)$.

PROOF. Let $x_0 \in (a, b)$. Let w be a solution of the differential equation (b) with $w(x_0) = w_0$, $w'(x_0) = w_0'$, $w''(x_0) = w_0''$ and let at least one of the numbers w_0, w_0', w_0'' be non-zero. Choose the initial conditions for the desired solutions y_1, y_2 of (a) so that

$$\begin{vmatrix} y_1, & y_2 \\ y_1', & y_2' \end{vmatrix}(x_0) = w_0, \quad \begin{vmatrix} y_1, & y_3 \\ y_1', & y_2' \end{vmatrix}'(x_0) = w_0',$$

$$\begin{vmatrix} y_1, & y_2 \\ y_1', & y_2' \end{vmatrix}''(x_0) = w_0''$$

and so that at least one of the numbers $y_1(x_0)$, $y_1'(x_0)$, $y_1''(x_0)$ and at least one of the numbers $y_2(x_0)$, $y_2'(x_0)$, $y_2''(x_0)$ is non-zero. This is always possible. Thus we obtain a function $w = y_1 y_2' - y_1' y_2$ satisfying the initial conditions $w(x_0) = w_0$, $w'(x_0) = w_0'$, $w''(x_0) = w_0''$. By Section 2, w is a solution of the differential equation (b). Thus we have determined y_1, y_2 and proved Lemma 1.1. □

COROLLARY 1.4. Let $w(x) \neq 0$ for $x \in I \subset (a, b)$ be a solution of the differential equation (b) and assume that y_1, y_2 are independent solutions of the differential equation (a), while $w = y_1 y_2' - y_1' y_2$. Then the function

$$y_3(x) = \int_{x_0}^{x} \frac{y_1(x) y_2(t) - y_1(t) y_2(x)}{w^2(t)}\, dt, \qquad x_0 \in I \tag{1.10}$$

is a solution of the differential equation (a) and satisfies $y_3(x_0) = y_3'(x_0) = 0$, $y_3''(x_0) \neq 0$.

PROOF. It is easy to verify that the set of solutions $y = c_1 y_1 + c_2 y_2$ satisfies for $x \in I$ the following equation of the form (c)

$$wy'' - w'y' + (w'' + 2Aw)y = 0 .$$

By termwise differentiation of the last equation we get the equation (a), because all its solutions are at the same time solutions of (a).

Consider the following non-homogeneous differential equation

$$wy'' - w'y' + (w'' + 2Aw)y = w(x_0)y_3''(x_0) . \tag{1.11}$$

Differentiating this term by term we again get the differential equation (a), hence all the solutions of (1.11) are solutions of (a).

By the method of variation of constants it is easy to see that the function (1.10) satisfies the differential equation (1.11) and thus also the differential equation (a), which was to be proved. □

REMARK 1.5. In the sequel, the set of solutions $y = c_1y_1 + c_2y_2$ will be referred to as the band of solutions of the differential equation (a) in the interval I. The formula (1.10) was originally derived by Sansone [125], who also showed that y_1, y_2, y_3 form a fundamental system of solutions of the differential equation (a) on I.

THEOREM 1.13. The differential equation (a) has at least one solution without zeros in the interval (a, b).

PROOF. Let y_1, y_2, y_3 be a fundamental system of solutions of the differential equation (a) satisfying

$$y_1(x_0) = y_1'(x_0) = 0,\ y_1''(x_0) = 1,\ y_2(x_0) = y_2''(x_0) = 0,$$
$$y_2'(x_0) = 1,\ y_3'(x_0) = y_3''(x_0) = 0,\ y_3(x_0) = 1,\ x_0 \in (a, b).$$

Let $x_0 < x_1 < x_2 < \ldots$ be an infinite sequence of numbers converging to the point b. The integral identity (1.6) implies that y_1 has no zero to the left of x_0.

Now construct a sequence $\{u_n\}_{n=1}^{\infty}$ of solutions of the differential equation (a), $u_n = c_1^n y_1 + c_2^n y_2 + c_3^n y_3$ with $u_n(x_n) = u_n'(x_n) = 0$, $u_n''(x_n) > 0$ and such that $u_n^2(x_0) + u_n'^2(x_0) + u_n''^2(x_0) = 1$. This is evidently possible. From the integral identity (1.6) for u_n it follows that u_n has no zero to the left of x_n and that $u_n(x) > 0$ for $x < x_n \in (a, b)$.

Consider the following number sequences:

$$\{u_n(x_0)\}_{n=1}^{\infty}, \quad \{u_n'(x_0)\}_{n=1}^{\infty}, \quad \{u_n''(x_0)\}_{n=1}^{\infty}. \tag{1.12}$$

Evidently, each of these sequences is bounded, so there exist convergent subsequences of these sequences (1.12). For simplicity, we denote them again by (1.12) and their limits by u_0, u_0', u_0''.

Let u be the solution of the differential equation (a) satisfying $u(x_0)=u_0$, $u'(x_0)=u_0'$, $u''(x_0)=u_0''$. This solution is non-trivial since all the numbers u_0, u_0', u_0'' cannot vanish simultaneously, because $\lim_{n\to\infty} [u_n^2(x_0)+u_n'^2(x_0)+u_n''^2(x_0)]=u_0^2+u_0'^2+u_0''^2=1$. The solution u_n may be expressed in the form $u_n=u_n''(x_0)y_1+u_n'(x_0)y_2+u_n(x_0)y_3$ and the solution $u=u_0''y_1+u_0'y_2+u_0y_3$. Hence $\lim_{n\to\infty} u_n(x)=u(x)$ for all $x\in(a, b)$ due to uniform convergence in every closed subinterval of (a, b).

We show now that $u(x)\neq 0$ for $x\in(a, b)$. The integral identity (1.6) for the solution u_n becomes

$$u_nu_n''-\frac{1}{2}u_n'^2+Au_n^2+\int_{x_0}^{x} bu_n^2\,dt=0,$$

i. e.

$$u_nu_n''-\frac{1}{2}u_n'^2+Au_n^2=\int_{x}^{x_n} bu_n^2\,dt.$$

It follows that the integral on the right-hand side exists for each $x\in(a, b)$ and for every positive integer n furthermore that its limit, as $n\to\infty$, is

$$uu''-\frac{1}{2}u'^2+Au^2=\int_{x}^{b} bu^2\,dt+k, \qquad k\geqq 0. \tag{1.13}$$

This is an integral identity for the solution u. From the integral identity (1.13) it follows that u has no zero in the interval (a, b). □

REMARK 1.6. Similarly it can be shown, that if $b(x)$ has the property (v), then the differential equation (b) has at least one solution without zeros in the interval (a, b). In the proof it is necessary to take into account that the $\{x_n\}_{n=1}^{\infty}$ converges to a as $n\to\infty$.

REMARK 1.7. If $a=-\infty$, $b=\infty$, then Theorem 1.13 implies that the differential equation (a), under the assumption (v), has at least one solution without zeros on the whole real line. Every third order differential equation with constant coefficients has a similar property.

Let $w\neq 0$ for $x\in(a, b)$ be a solution of the differential equation (b). Let y_1, y_2 be independent solutions of the differential equation (a) satisfying $w=y_1y_2'-y_1'y_2$. Such solutions exist by Lemma 1.1. Then the set of solutions $y=c_1y_1+c_2y_2$ satisfies the equation of the form (c) on the whole interval (a, b). If $a=-\infty$, this set will be called the band at the improper point $-\infty$.

COROLLARY 1.5. The differential equation (a) has a band of solutions that satisfy a differential equation of the form (c) on the whole interval (a, b), while $w(x)\neq 0$ for $x\in(a, b)$.

REMARK 1.8. Using bands, e. g. at $-\infty$, differential equations of the form (a) with certain properties can be constructed. For instance, differential equations of the form (a) whose adjoint equations have $w=e^x$ as a solution without zeros in $(-\infty, \infty)$ take the form

$$y'''+2Ay'+(A'+2A+A'+1)y=0\,,$$

where A' is a continuous function of x $(-\infty, \infty)$. The above equation can be obtained by termwise differentiation of the second order equation

$$e^x y''-e^x y'+(e^x+2A\,e^x)y=0\,.$$

COROLLARY 1.6. Let A, A', b be continuous functions of $x\in(-\infty, \infty)$ and let b have the property (v) in $(-\infty, \infty)$ Then every solution of the differential equation (a) having a zero has infinitely many zeros in (α, ∞) if and only if every solution in the band at $-\infty$ has infinitely many zeros in (α, ∞), $\alpha>-\infty$.

REMARK 1.9. The concept of a band can be defined for the differential equation (b) in the same way that we have introduced the concept of a band at x_0, $a<x_0<b$, for the differential equation (a). Assuming the function $b=b(x)$ to have the property (v), all the

properties of bands of solutions and their consequences for solutions of the differential equation (a) remain in force also for solutions of the differential equation (b), but on the left of the point x_0 at which the band is considered.

8. *Weakening of Property (v) for the Laguerre Invariant*

Let $b(x)\equiv 0$ for $x\in(a, b)$. Then the differential equation (a) becomes self-adjoint and its fundamental set of solutions is u^2, v^2, uv, where u and v form a fundamental system of solutions of the second order differential equation (1.4). Obviously, every zero of u or v is a double zero of the corresponding solution u^2 or v^2 of the differential equation (1.3).

COROLLARY 1.7. The self-adjoint third order differential equation (1.3) has at least one solution with no zero in the interval (a, b), namely the function u^2+v^2.

LEMMA 1.2 Let $\bar{u}$, $\bar{v}$ be solutions of the differential equation (a) with $\bar{u}(\alpha_1)=\bar{u}(\alpha_2)=0$, where $\alpha_1<\alpha_2\in(a, b)$ are two consecutive zeros of the solution $\bar{u}$ and $\bar{u}(x)>0$ for $x\in(\alpha_1, \alpha_2)$. Let $\bar{v}(x)>0$ for $\alpha_1\leqq x\leqq\alpha_2$. Then there exist a constant $\lambda>0$ and $\tau\in(\alpha_1, \alpha_2)$ such that the solution $\bar{v}-\lambda\bar{u}$ has a double zero at τ.

PROOF. Let $\bar{u}$, $\bar{v}$ meet the hypotheses of Lemma 1.2. Then evidently

$$\left(\frac{\bar{u}}{\bar{v}}\right)'=\frac{\bar{v}\bar{u}'-\bar{u}\bar{v}'}{\bar{v}^2}.$$

Integration of the last equality from α_1 to α_2 gives

$$0=\int_{\alpha_1}^{\alpha_2}\frac{\bar{v}\bar{u}'-\bar{u}\bar{v}'}{\bar{v}^2}\,dt.$$

This can only hold if the function $\bar{v}\bar{u}'-\bar{u}\bar{v}'$ has at least one zero between α_1 and α_2 at which it changes its sign. Denote such a zero by τ. Then, at τ, the equations

$$c_1\bar{u}(\tau)-c_2\bar{v}(\tau)=0\ ,$$
$$c_1\bar{u}'(\tau)+c_2\bar{v}'(\tau)=0$$

have a non-trivial solution for c_1, c_2, whence the assertion follows by putting $\lambda=\bar{u}(\tau)/\bar{v}(\tau)$. □

Suppose, further, that the Laguerre invariant has the property

$$b(x)\geqq 0 \text{ for } x\in(a,b)\ . \tag{v$_1$}$$

The following results are valid [Rovder, 121].

THEOREM 1.14. If there is at least one solution of the differential equation (a) with infinitely many zeros in (α, b), $a<\alpha<b$, then every solution of (a) having a zero has infinitely many zeros in (a, b).

PROOF. Let y be a solution of the differential equation (a) having infinitely many zeros in (α, b) and assume that $y(x_0)=0$, $y'(x_0)>0$. We show first that if $\bar{y}$ is a solution of the differential equation (a) with $\bar{y}(x_0)=\bar{y}'(x_0)=0$, $\bar{y}''(x_0)\neq 0$, then it has infinitely many zeros in (α, b). Suppose the contrary, i.e. let there exist $x_1>x_0\in(a,b)$ such that $\bar{y}(x)>0$ for $x>x_1$. Let $\alpha_1, \beta_1\in(x_1, b)$ be two neighbouring zeros of the solution y. By Lemma 1.2, there is a number $c>0$ such that the solution $v=\bar{y}-cy$ has a double zero at some $\tau\in(\alpha_1,\beta_1)$. The solution v has thus the property $v(x_0)=0$, $v'(x_0)\neq 0$, $v(\tau)=v'(\tau)=0$, $\tau>x_0$.

The integral identity (1.6) for the solution v takes the form

$$vv''-\frac{1}{2}v'^2+Av^2+\int_\tau^x bv^2\,dt=0\ .$$

At $x_0<\tau$ we have

$$-\frac{1}{2}v'^2(x_0)=\int_{x_0}^\tau bv^2\,dt\ ,$$

which is a contradiction, since $v'(x_0)\neq 0$. Therefore, $\bar{y}$ has infinitely many zeros in (a, b).

It can be proved similarly that if $y(x_0)=y'(x_0)=0$, $y''(x_0)\neq 0$ and $\bar{y}(x_0)=0$, $\bar{y}'(x_0)\neq 0$, then $\bar{y}$ has infinitely many zeros in (α, b). If y and $\bar{y}$ have simple zeros at x_0, the above reasoning implies that there is a solution u of the differential equation (a) with $u(x_0)=u'(x_0)=0$, $u''(x_0)\neq 0$ which has infinitely many zeros in (α, b) whenever y has infinitely such zeros. Thus, $\bar{y}$ has infinitely many zeros in (α, b).

Now let y have infinitely many zeros in (α, b) and let $\bar{y}$ satisfy $\bar{y}(x_0)=0$, $x_0 \in (\alpha, b)$. If x_0 coincides with some zero of the solution y, then $\bar{y}$ has infinitely many zeros in (α, b). Let $x_1 \neq x_0$ be a zero of the solution y. Then there exists a solution v of the differential equation (a) with $v(x_1)=v(x_0)=0$. From what has already been shown, this solution has infinitely many zeros in (α, b), and hence $\bar{y}$ has the same property. □

REMARK 1.10. Theorem 1.14 does not imply separation of zeros of bands of solutions of the differential equation (a).

LEMMA 1.3. Assume that the coefficient b in the differential equation (a) has the property (v_1). Then there exists at least one solution of the differential equation (a) which is non-negative in the interval (a, b).

The proof of Lemma 1.3 is similar to that of Teorem 1.13 except the last part, which is omitted. It is necessary to realize that because of the integral identity (1.6) no solution of (a) with a double zero at $x_n \in (a, b)$ has a simple zero to the left of x_n.

THEOREM 1.15. If the function b has the property (v_1), then there exists a solution of the differential equation (a) without zeros in (a, b).

PROOF. Suppose that every solution of the differential equation (a) having a zero has infinitely many zeros in (α, b), $a<\alpha<b$. If $b(x)\equiv 0$ on some interval with the right endpoint b, then the equation (a) is self-adjoint there, and hence there exists a solution without zeros in that interval. By Theorem 1.14, this solution is non-zero on the whole interval (a, b), else it would necessarily have infinitely many zeros in (α, b).

Let $b(x)\not\equiv 0$ in any subinterval of (a, b) with the right hand end at b and let $y(x)$ be a solution of the differential equation (a), non-negative in (a, b). Such a solution exists, by Lemma 1.3. Let $\bar{x}\in(a, b)$ be its double zero. It follows from the integral identity (1.13) that this is impossible.

Suppose now that every solution of the differential equation (a) has finitely many zeros in (α, b), $a<\alpha<b$. Let y be a solution of the differential equation (a) with $y(x)\geqq 0$ for $x\in(a, b)$, such a solution exists by Lemma 1.3 and let $x_0\in(a, b)$ be its last double zero. Let $u(x)$ be a solution of the differential equation (a) satisfying the following conditions:

$$u(x_1)=y(x_1)\,,\quad u'(x_1)=y'(x_1)\,,\quad u''(x_1)>y''(x_1)\,,$$
$$x_1>x_0\in(a, b)\,. \tag{1.14}$$

The function $z=u-y$ then satisfies at x_1 the conditions $z(x_1)=z'(x_1)=0$, $z''(x_1)>0$. It follows from (1.6) that $z(x)\geqq 0$ for $x\in(a, x_1)$, and therefore $u(x)\geqq y(x)\geqq 0$ in the interval $(a, x_1\rangle$. Conditions (1.14) imply that u and y are linearly independent and thus cannot have double zeros in common. Therefore $u(x)>0$ in $(a, x_1\rangle$.

Now we show that z has no zero in (x_1, b). Suppose the contrary, i.e. let z have a zero in (x_1, b). Then, by Lemma 1.2, there is a constant $c\neq 0$ such that the solution $cz-y$ of (a) has a double zero at some $\tau_2>x_1$. Also, the function $cz-y$ evidently has a simple zero at $\tau_1<x_1$. Hence $\tau_1<\tau_2$. However, in view of (1.6), this is not possible because a solution having a double zero cannot have a simple zero to the left of the double one. Thus we must have $z(x)>0$ in (x_1, b), that is, $u(x)>y(x)>0$ in (x_1, b). Therefore $u(x)>0$ on the whole interval (a, b), and the theorem is proved. □

§2. OSCILLATORY PROPERTIES OF SOLUTIONS OF THE DIFFERENTIAL EQUATION (a)

1. *Basic Definitions*

The differential equation (a) is said to be *disconjugate* in (a, b) if every non-trivial solution of (a) has at most two zeros (multiplicity being counted) in the interval (a, b).

A non-trivial solution of the differential equation (a) is called *oscillatory* in (a, b) if b is a limit point of zeros of that solution. In the contrary case we say that the solution of (a) is *non-oscillatory* in (a, b).

The differential equation (a) is said to be oscillatory in the interval (a, b) if it has at least one oscillatory solution in (a, b). If the differential equation (a) has no oscillatory solution in (a, b), we say that it is non-oscillatory in (a, b).

If the Laguerre invariant $b = b(x)$ has the property (v) in (a, b), then Theorem 1.3 immediately implies the following proposition.

A necessary condition for the differential equation (a) to be oscillatory in (a, b) is the existence of α, $a < \alpha < b$, such that every solution of the differential equation (b) having a double zero in (α, b) has another zero in (a, b), to the right of the double one.

The notions of disconjugateness, oscillatoricity and non-oscillatoricity may be generalized and defined analogously for a linear differential equation of order n $(n = 2, 3, 4, \ldots)$.

A linear homogeneous differential equation of order n is called *disconjugate* in (a, b) if each of its non-trivial solutions has at most $n - 1$ zeros (multiplicity being counted) in (a, b).

A non-trivial solution of a linear differential equation of order n is said to be *oscillatory* in (a, b) if b is a limit point of its zeros. In the contrary case we say that the solution is *non-oscillatory* in (a, b).

A linear homogeneous differential equation of order n is called oscillatory in (a, b) if it has at least one oscillatory solution in (a, b). Otherwise we say that the equation is non-oscillatory in (a, b).

2. *Sufficient Conditions for the Differential Equation (a) to Be Disconjugate*

THEOREM 2.1. Let $b = b(x)$ have the property (v). Also, assume that $A(x) \leqq 0$ for $x \in (a, b)$ and $|A(x)| \geqq \int_\alpha^x b(t)\, dt$ for every $\alpha \leqq x \in (a, b)$. Then the differential equation (a) is disconjugate in the interval (a, b).

PROOF. The theorem will be established using Theorem 1.11 once we show that no solution of (a) with a double zero has another zero. The assertion then follows from the properties of bands.

Let $a<\alpha<b$. Let y be a solution of the differential equation (a) with $y(\alpha)=y'(\alpha)=0$, $y''(\alpha)>0$.

The integral identity (1.6) for the solution y takes the form

$$yy''-\frac{1}{2}y'^2+Ay^2+\int_\alpha^x by^2\,dt=0\,.$$

Hence it follows directly that y has no zero to the left of α.

Integrate the last identity from α to x. We obtain

$$y(x)y'(x)=\frac{3}{2}\int_\alpha^x y'^2(t)\,dt+$$
$$+\int_\alpha^x\left[|A(t)|y^2(t)-\int_\alpha^t b(s)y^2(s)\,ds\right]dt\,.$$

Suppose the assertion of the theorem is false, and let $x_1>\alpha\in(a, b)$ be the first zero of the function $y(x)$. The hypotheses evidently imply that

$$|A(t)|y^2(t)-\int_\alpha^t b(s)y^2(s)\,ds\geqq$$
$$\geqq y^2(t)\left[|A(t)|-\int_\alpha^t b(s)\,ds\right]\geqq 0\,,$$

and the above inequality for $y(x)y'(x)$ gives rise to a contradiction at x_1. Thus the theorem is proved. □

REMARK 2.1. Theorem 2.1 was originally stated and proved by Sansone [125], but for a closed interval $\langle a, b\rangle$.

COROLLARY 2.1. The assertion of Theorem 2.1 remains true also for the differential equation (b).

The proof follows from Theorem 1.3 and Remark 1.9.

THEOREM 2.2. Assume that $A(x)\leqq 0$, $b(x)$ has the property (v) and $b(x)-A'(x)\leqq 0$ for $x\in(a, b)$. Then the differential equation (a) is disconjugate in (a, b).

PROOF. Let $a<\alpha<b$. Let y be a solution of the differential equation (a) with $y(\alpha)=y'(\alpha)=0$, $y''(\alpha)>0$. We have to show that y has no other zero in (a, b). From the integral identity (1.6) it follows that y has no zero to the left of α in (a, b). It remains to show that $y(x)>0$ for $x>\alpha$. Evidently,

$$(yy')' = yy'' + y'^2.$$

Substitute for y'' in the above relation the expression implied by the integral identity (1.7) for the solution y. This gives

$$(yy')' = y''(\alpha)y - 2Ay^2 - y\int_v^x (b - A')y\, dt + y'^2.$$

Assuming that $y_1 > \alpha \in (a, b)$ is the first zero of the solution y to the right of α, we get a contradiction after integrating the last equality from α to x_1. Thus the theorem is proved. □

REMARK 2.2. The assumptions $A(x) \leqq 0$, $A'(x) + b(x) \leqq 0$ for $x \in (a, b)$ and the assumption that b has the property (v) for $x \in (a, b)$ provide also a sufficient condition for the differential equation (a) to be disconjugate in (a, b).

Substitute $w = \exp(-x)$ in the differential equation (c). We get

$$e^{-x}y'' + e^{-x}y' + e^{-x}(2A+1)y = 0. \tag{c'}$$

Differentiating the last equation term by term after cancelling $\exp(-x)$, yields the equation

$$y''' + 2Ay' + (2A' - 2A - 1)y = 0. \tag{2.1}$$

In virtue of Remark 1.8, $w = \exp(-x)$ is a solution without zeros in (a, b) of the differential equation adjoint to (2.1).

LEMMA 2.1. Let $2A(x) + 1 \leqq 0$ for $x \in (a, b)$. Then every solution of the differential equation (c′) has at most one zero.

PROOF. The differential equation (c′) may be written in the form $y'' + y' = -(2A+1)y$.

Using the method of variation of constants, it is easy to see that every solution of (c′) satisfies

$$y = c_1 + c_2 e^{-x} + \int_\alpha^x \frac{e^{-x} - e^{-t}}{e^{-t}} [2A(t) + 1]y(t)\, dt.$$

From the last equality and the hypotheses of Lemma 2.1, the above assertion follows. □

LEMMA 2.2. Let $2A(x)+1 \leqq 0$ and $2A(x)+1-A'(x) \leqq 0$ for $x \in (a, b)$ and assume that equality in the latter formula does not hold in any subinterval of (a, b). Then the differential equation (2.1) is disconjugate in (a, b).

To prove the lemma, it is necessary to show that solutions of the differential equation (2.1) having a double zero at some point of the interval (a, b) have no other zero in (a, b). The assertion follows from Lemma 2.1 and from the properties of bands of solutions of the differential equation (2.1). In fact, the set of solutions of the differential equation (c′) is a subset of the solutions of the differential equation (2.1) (the band at a) and, at the same time, the differential equation (c′) is disconjugate, i.e., each of its solutions has at most one zero. □

THEOREM 2.3. Let the hypotheses of Lemma 2.2 be met, let $0 \leqq b(x) \leqq A'(x) - 2A(x) - 1$, and assume that b has the property (v) in (a, b). Then the differential equation (a) is disconjugate in (a, b).

PROOF. Again, we have to show that the solutions of (a) having a double zero have no other zero in (a, b) to the right of the double one.

The differential equation (a) may be written in the form

$$y''' + 2Ay' + (2A' - 2A - 1)y = (A' - 2A - 1 - b)y \,.$$

Using the method of variation of constants, we easily verify that every solution y of (a) can be expressed in the following form:

$$y(x) = z(x) + \int_a^x [A'(t) - 2A(t) - b(t) - 1] \,.$$
$$. \, W(x, t)y(t)\, dt \,, \tag{2.2}$$

where z is a solution of (2.1), satisfying the same initial conditions as y, and

$$W(x,t)=\begin{vmatrix} z_1(x), & z_2(x), & z_3(x) \\ z_1(t), & z_2(t), & z_3(t) \\ z_1'(t), & z_2'(t), & z_3'(t) \end{vmatrix},$$

where z_1, z_2, z_3 is a fundamental system of solutions of (2.1) whose wronskian equals 1. The function $W(x, t)$, for a fixed t, is a solution of the differential equation (2.1) with a double zero at $x=t$ and hence, by Lemma 2.2, $W(x, t) \geqq 0$ for $t \geqq x$. The assertion of the theorem follows from (2.2) and Lemma 2.2.

3. *Sufficient Conditions for Oscillatoricity of Solutions of the Differential Equation (a)*

Sansone [125] has proved the following comparison theorem for the differential equation (a) and the self-adjoint third order equation

$$z''' + 2A_1 z' + A_1' z = 0\,, \tag{2.3}$$

where $A_1 = A_1(x)$ and $A_1' = A_1'(x)$ are continuous functions of $x \in (a, b)$.

THEOREM 2.4. Let $b(x) \geqq 0$ and $A(x) \geqq A_1(x)$ for $x \in (a, b)$. Let z be a non-trivial solution of the differential equation (2.3) with $z(\alpha) = z'(\alpha) = 0$, $z(\beta) = z'(\beta) = 0$, where $\alpha < \beta \in (a, b)$ are two consecutive zeros of the solution z. Further, assume that $x_0 \leqq \alpha \in (a, b)$ and that y is a solution of the differential equation (a) with

$$y(x_0)y''(x_0) - \frac{1}{2}y'^2(x_0) + A(x_0)y^2(x_0) \leqq 0\,. \tag{2.4}$$

Then the solution y has at least one zero in the interval (α, β) except for the case when $A \equiv A_1$, $b \equiv 0$ for $\alpha \leqq x \leqq \beta$ and y is linearly dependent on z.

PROOF. Suppose the contrary, i.e. that y has no zero in (α, β) and let $y = Y^2(x)$, where $Y(x) > 0$ for $\alpha < x < \beta$. Moreover, let

$$y(x_0)y''(x_0) - \frac{1}{2}y'^2(x_0) + A(x_0)y^2(x_0) = -K^2, \quad K \geqq 0\,.$$

The integral identity (1.6) for y then becomes

$$yy''-\frac{1}{2}y'^2+Ay^2+\int_{x_0}^{x} by^2\,dt=-K^2.$$

We deduce the following relation for Y:

$$Y''+\frac{1}{2}AY=-\frac{1}{2Y^3}\left[K^2+\int_{x_0}^{x} by^2\,dt\right],\quad \alpha<x<\beta\,. \tag{2.5}$$

Let $z(x)>0$ for $\alpha<x<\beta$ and let $z(x)=Z^2(x)$, that is, $Z(x)=\sqrt{z(x)}>0$ for $\alpha<x<\beta$.

According to Section 2 of §1, Z satisfies the second order differential equation

$$Z''+\frac{1}{2}A_1Z=0\,. \tag{2.6}$$

Apply to Y and Z the so-called Picone identity for solutions of the second order equations (Sansone [124]). This gives

$$\begin{aligned}
&\frac{d}{dx}\left[\frac{Z}{Y}(Z'Y-ZY')\right]=\frac{d}{dx}\left[ZZ'-Z^2\frac{Y'}{Y}\right]=\\
&=Z'^2+ZZ''-2ZZ'\frac{Y'}{Y}-Z^2\frac{Y''}{Y}+Z^2\frac{Y'^2}{Y^2}=\\
&=\left(Z'-Z\frac{Y'}{Y}\right)^2-\frac{1}{2}A_1Z^2+\\
&+\frac{Z^2}{Y}\left[\frac{1}{2}AY+\frac{1}{2Y^3}\left(K^2+\int_{x_0}^{x} by^2\,dt\right)\right]=\\
&=\left(Z'-Z\frac{Y'}{Y}\right)^2+\frac{1}{2}(A-A_1)Z^2+\frac{Z^2}{2Y^4}\left(K^2+\int_{x_0}^{x} by^2\,dt\right).
\end{aligned} \tag{2.7}$$

Taking the integral in (2.7) from $\alpha+\varepsilon$ to $\beta-\varepsilon$ with $\varepsilon>0$ such that $\alpha+\varepsilon<\beta-\varepsilon$, we get

$$\begin{aligned}
&\left[ZZ'-Z^2\frac{Y'}{Y}\right]_{\alpha+\varepsilon}^{\beta-\varepsilon}=\int_{\alpha+\varepsilon}^{\beta-\varepsilon}\left(Z'-Z\frac{Y'}{Y}\right)^2 dt+\\
&+\frac{1}{2}\int_{\alpha+\varepsilon}^{\beta-\varepsilon}(A-A_1)Z^2\,dt+\frac{1}{2}\int_{\alpha+\varepsilon}^{\beta-\varepsilon}\frac{Z^2}{Y^4}\left(K^2+\int_{x_0}^{t} by^2\,ds\right)dt\,.
\end{aligned} \tag{2.8}$$

If y, and hence also Y, is non-zero both at α and β, and if $Z(\alpha)=Z(\beta)=0$, we obtain

$$0=\int_\alpha^\beta \left(Z'-Z\frac{Y'}{Y}\right)^2 dt+\frac{1}{2}\int_\alpha^\beta (A-A_1)Z^2\,dt+$$

$$+\frac{1}{2}\int_\alpha^\beta \frac{Z^2}{Y^4}\left(K^2+\int_{x_0}^t by^2\,ds\right)dt\,, \tag{2.9}$$

which is a contradiction, therefore both Y and y have at least one zero in (α, β).

If $y(\alpha)=0$, we get from (2.8) that

$$\lim_{x\to\alpha^+}\left[ZZ'-Z^2\frac{Y'}{Y}\right]=\frac{1}{2}\lim_{x\to\alpha^+}\left[z'-z\frac{y'}{y}\right]=0\,.$$

The situation at $\beta-$ is similar. Thus, a contradiction arises again from (2.8) or (2.9).

If the differential equations (a) and (2.3) are identical for $\alpha\leqq x\leqq\beta$ and y and z are dependent, then $A\equiv A_1$, $K=0$, $b\equiv 0$ for $\alpha\leqq x\leqq\beta$ and $Z'Y-ZY'=0$, therefore $Y=cZ$, where c is a suitable constant. Thus the theorem is proved. □

COROLLARY 2.2. Let the hypotheses of Theorem 2.4 be satisfied and, moreover, assume that the differential equation (2.6) has an oscillatory solution in (α, b), $a<\alpha<b$, i.e. at least one solution has infinitely many zeros in (α, b). Then every solution of the differential equation (a) with the property (2.4), $a<x_0<b$, oscillates in (α, b).

COROLLARY 2.3. Let the function b have the property (v) and let the second order differential equation (1.4) have oscillatory solutions in (α, b), $a<\alpha<b$. Then the differential equation (a) is oscillatory in (a, b), i.e. every solution of (a) having a zero (or satisfying (2.4)) is oscillatory in (a, b).

Besides the differential equation (a), let us consider the following differential equation

$$z'''+2A_1z'+(A_1'+b_1)z=0\,, \tag{a_1}$$

where $A_1=A_1(x)$, $A_1'=A_1'(x)$, $b_1=b_1(x)$ are continuous functions of $x\in(a, b)$.

LEMMA 2.3. Let y be any solution of the differential equation (a). Then it can be expressed in the form

$$y(x) = z(x) + \int_{x_0}^{x} [2A_1(t) - 2A(t)]\, y'(t)\, W(x, t)\, dt +$$

$$+ \int_{x_0}^{x} [b_1(t) - b(t) + A_1'(t) - A'(t)]\, y(t)\, W(x, t)\, dt\,, \tag{2.10}$$

where z is the solution of the differential equation (a_1), which has the same initial values at $x_0 \in (a, b)$ as y, $W(x, t)$ is a function of the form

$$W(x, t) = \begin{vmatrix} z_1(x), & z_2(x), & z_3(x) \\ z_1(t), & z_2(t), & z_3(t) \\ z_1'(t), & z_2'(t), & z_3'(t) \end{vmatrix}$$

and z_1, z_2, z_3 form a fundamental set of solutions of the differential equation (a_1) with wronskian equal to 1 on (a, b).

The proof will be carried out by applying the method of variation of constants to the differential equation (a) written in the form

$$y''' + 2A_1 y' + (A_1' + b_1) y =$$
$$= (2A_1 - 2A) y' + (b_1 - b + A_1' - A') y\,.$$

Put $A_1 = A$ in the differential equation (a_1) and denote the result by (a_2),

$$z''' + 2Az' + (A' + b_1) z = 0\,. \tag{a_2}$$

We now prove the following comparison theorem.

THEOREM 2.5. Let the function b have the property (v) in (a, b) and let $b(x) \leqq b_1(x)$ in (a, b). Then the following assertion holds:

If the differential equation (a) is oscillatory in (a, b), then so is (a_2). If y and z are solutions of (a) and (a_2) with $y(x_0) = y'(x_0) = 0$, $y''(x_0) \neq 0$, $z(x_0) = z'(x_0) = 0$, $z''(x_0) \neq 0$, $a < x_0 < b$, then the first zero of z to the right of x_0 in (a, b) is not farther than the first zero of y to the right of x_0 in (a, b).

PROOF. Suppose that the differential equation (a) is oscillatory in (a, b). It is sufficient to show that every solution of the differential equation (a_2) having a double zero at x_0, $a < x_0 < b$, has another zero on the right of x_0 not beyond a zero of the solution of (a) having a double zero at x_0.

Let y be a solution of the differential equation (a) with $y(x_0)=y'(x_0)=0$, $y''(x_0)>0$, $a<x_0<b$, let z be the solution of the differential equation (a_2) satisfying $z(x_0)=z'(x_0)=0$, $z''(x_0)=y''(x_0)$, and let $x_1>x_0$ be the first zero of y in (a, b). The differential equation (a_2) can be rewritten in the form

$$z'''+2Az'+(A'+b)z=-(b_1-b)z\,.$$

By Lemma 2.3,

$$z(x)=y(x)-\int_{x_0}^{x}[b_1(t)-b(t)]z(t)\,W(x,t)\,dt\,, \tag{2.11}$$

where the function $W(x, t)$ is constructed using a fundamental system of (a) whose wronskian equals 1. For a fixed t, $W(x, t)$ satisfies the differential equation (a) and has a double zero at t, and therefore (by Theorem 1.9) $W(x, t)\geqq 0$ for $t\leqq x\in\langle x_0, x_1\rangle$. The assertion of Theorem 2.5 then follows from (2.11). □

Substitute in the differential equation (c) $w=\exp(nx)$, where n is a positive integer. The following equation will result:

$$e^{nx}y''-n\,e^{nx}y'+e^{nx}(n^2+2A)y=0\,. \tag{2.12}$$

Differentiating the equation (2.12) term by term and cancelling $\exp(nx)$ yields the differential equation

$$y'''+2Ay'+(A'+A'+n^3+2An)y=0\,. \tag{a'}$$

COROLLARY 2.4. Let $A(x)>-k$, $A'(x)>-k$, $k>0$ and let n be such that the differential equation (2.12) (with $\exp(nx)$ cancelled) oscillates in (a, b) and $b(x)\geqq n^3+2An+A'>0$ for $x\in(a, b)$. Then the differential equation (a) also oscillates in (a, b).

The proof follows from Theorem 2.5 by comparison of equations (a) and (a').

COROLLARY 2.5. If a solution of the differential equation (a_2) with a double zero at x_0, $a<x_0<b$, has no other zero to the right of x_0 in (a, b), then no solution of (a) having a double zero at some point of the interval $\langle x_0, b)$ has other zeros.

4. *Further Conditions Concerning Oscillatoricity or Non-oscillatoricity of Solutions of the Differential Equation (a)*

Let (c) represent the differential equation of a band at x_0, $a<x_0<b$, and let $w(x)>0$ for $x>x_0 \in (a, b)$. By the substitution $y=\sqrt{w}\, u$, the differential equation (c) is transformed into

$$u''+\left[\frac{3w''}{2w}-\frac{3w'^2}{4w^2}+2A\right]u=0\,. \tag{2.13}$$

The integral identity (1.8) for w gives

$$\frac{3}{2}\frac{w''}{w}-\frac{3}{4}\frac{w'^2}{w^2}+2A=\frac{1}{2}A+\frac{3}{2w^2}\int_{x_0}^{x} bw^2\,dt,\quad x>x_0.$$

Thus the differential equation (2.13) may be written in the form

$$u''+\left[\frac{1}{2}A+\frac{3}{2w^2}\int_{x_0}^{x} bw^2\,dt\right]u=0\,. \tag{2.14}$$

An immediate consequence of the above reasoning is the following theorem.

THEOREM 2.6. Let the function b have the property (v). A necessary and sufficient condition for the differential equation (a) to be oscillatory in (a, b) is that the solutions of the differential equation (2.14) should oscillate in (α, b), $a<\alpha<b$.

REMARK 2.3. If $b(x)\equiv 0$ for $x\in(a, b)$, then the equation (2.14) reduces to (1.4) and the assertion of Theorem 2.5 remains true.

THEOREM 2.7. Let $A(x)\leqq 0$ and let b have the property (v) in (a, b). If the solutions of the differential equation

$$v''+\left[\frac{1}{2}A+\frac{3}{2}\int_{x_0}^{x} b\,dt\right]v=0 \tag{2.15}$$

are non-oscillatory in (a, b), that is, each of them has only finitely many zeros in (α, b), $x_0\leqq\alpha<b$, then the differential equation (a) is also non-oscillatory in (a, b).

PROOF. Let w be a solution of the differential equation (b) with $w(x_0)=w'(x_0)=0$, $w''(x_0)>0$, $a<x_0<b$. In virtue of Remark 1.3 we have $w(x)>0$, $w'(x)>0$ for $x>x_0$. The coefficients of the differential equations (2.14) and (2.15) satisfy

$$\frac{1}{2}A(x)+\frac{3}{2w^2(x)}\int_{x_0}^{x}b(t)w^2(t)\,dt=$$

$$=\frac{1}{2}A(x)+\frac{3}{2}\frac{w^2(\xi)}{w^2(x)}\int_{x_0}^{x}b(t)\,dt\leqq$$

$$\leqq\frac{1}{2}A(x)+\frac{3}{2}\int_{x_0}^{x}b(t)\,dt,\qquad a<\xi<x\,.$$

From the Sturm comparison theorem for the second order equations (Sansone [124]) (2.14) and (2.15) it follows that if the solutions of (2.15) are non-oscillatory in (x_0, b), then so are the solutions of (2.14), and by Theorem 2.6, the differential equation (a) is non-oscillatory in (a, b). Thus the theorem is proved. □

THEOREM 2.8. Let $A(x)\leqq 0$ and let $b(x)$ have the property (v) in the interval (a, b). Then a necessary condition for the differential equation (a) to be oscillatory in (a, b) is the oscillatoricity of the solutions of (2.15) in (x_0, b), $a<x_0<b$.

The proof follows again from comparison of the differential equations (2.14) and (2.15).

REMARK 2.4. Theorem 2.5 and Corollary 2.5 imply that every solution of the differential equation (a) having a zero has, under the hypotheses of Corollary 2.5, finitely many zeros to the right of that point.

THEOREM 2.9. Let f be a positive function having its third derivative f''' continuous in (a, b) and let $f'''+2Af'+A'f\geqq 0$ for $x\in(a, b)$, while equality does not hold in any subinterval of (a, b). Further, assume that

$$b(x)\geqq\frac{f'''(x)+2A(x)f'(x)+A'(x)f(x)}{f(x)}$$

for $x\in(a, b)$ and that the second order differential equation

$$u''+\left[\frac{3}{2}\frac{f''}{f}-\frac{3}{4}\frac{f'^2}{f^2}+2A\right]u=0 \tag{2.16}$$

has oscillatory solutions in (a, b). Then the differential equation (a) is oscillatory in (a, b). If

$$b(x)\leqq\frac{f'''(x)+2A(x)f'(x)+A'(x)f(x)}{f(x)}$$

and the solutions of (2.16) do not oscillate in (a, b), then the differential equation (a) is also non-oscillatory in (a, b).

The proof follows from comparison of the differential equation (a) with the differential equation

$$v'''+2Av'+\left[A'+\frac{f'''+2Af'+A'f}{f}\right]v=0\,, \tag{2.17}$$

which is obtained by differentiating the second order equation

$$fv''-f'v'+[f''+2Af]\,v=0\,. \tag{2.18}$$

By substitution $v=\sqrt{f}\,u$, the differential equation (2.18) becomes the differential equation (2.16). If the equation (2.16) is oscillatory in (a, b), then so is also (2.18). Then from the properties of bands it follows that the equation (2.17) is oscillatory and the comparison theorem implies that the equation (a) is also oscillatory. If, conversely, the differential equation (2.16) is non-oscillatory in (a, b), then so is the differential equation (a). □

COROLLARY 2.6. Let $A(x)\equiv 0$ for $x\in(a, b)$. Then the differential equation

$$y'''+by=0 \tag{2.19}$$

is oscillatory in (a, b) if $b(x)\geqq f'''(x)/f(x)\geqq 0$ with $f'''(x)/f(x)\not\equiv 0$ in any subinterval and if the second order differential equation

$$u''+\left[\frac{3}{2}\frac{f''}{f}-\frac{3}{4}\frac{f'^2}{f^2}\right]u=0 \tag{2.20}$$

is also oscillatory.

COROLLARY 2.7. If $f(x)=x^n$, n being a real number, $x\in(a, b)$, $a>0$, then the differential equation (2.20) has the form

$$u''+\frac{3(n^2-2n)}{4x^2}u=0 \tag{2.21}$$

and

$$\frac{f'''}{f}=\frac{n(n-1)(n-2)}{x^3}.$$

If $3(n^2-2n)=1$, that is, $n=1\pm 2/\sqrt{3}$, then the solutions of the differential equation (2.21) do not oscillate, and the solutions of (2.19) do not oscillate if $b(x)\leqq 2/3\sqrt{3}x^{-3}$.

If $3(n^2-2n)=1+\delta$, $\delta>0$, that is, $n=1\pm\sqrt{(4/3+3^{-1}\delta)}$, then the differential equation (2.21) has oscillatory solutions, and the differential equation (2.19) is oscillatory if

$$b(x)\geqq\frac{f'''}{f}=\frac{1}{3}(1+\delta)\sqrt{\frac{1}{3}(4+\delta)}\frac{1}{x^3}.$$

REMARK 2.5. The above conditions concerning the binomial equation of the third order were also established independently by Hanan [61].

5. *Relation between Solutions without Zeros and Oscillatoricity of the Differential Equation* (a)

In Theorem 1.13 we have proved that if the function b has the property (v) in (a, b), then the differential equation (a) has at least one solution without zeros in (a, b). The solution constructed in the proof of the theorem satisfies the integral identity (1.13).

LEMMA 2.4. Suppose $A(x)\leqq 0$ and let b have the property (v) for $x\in(a, b)$. Let y (z) be a solution of the differential equation (a) ((b)) satisfying $y(x_0)=y'(x_0)=0$, $y''(x_0)\neq 0$, $(z(x_0)=z'(x_0)=0$, $z''(x_0)\neq 0)$, $a<x_0<b$. Then both y (z) and its first derivative have no zeros to the left (right) of x_0.

LEMMA 2.5. Let $A(x) \leqq 0$, $A'(x)+b(x) \geqq 0$ and let b have the property (v) for $x \in (a, b)$. Let y (z) be a solution of the differential equation (a) ((b)) with $y(x_0)=y''(x_0)=0$, $y'(x_0) \neq$, $(z(x_0)=z''(x_0)=0$, $z'(x_0) \neq 0)$, $a < x_0 < b$. Then neither y (z) nor its first derivative has any zero to the left (right) of x_0.

The proof of Lemmas 2.4 and 2.5 for a solution z of the differential equation (b) is Theorem 1.7 and Remark 1.3. For a solution y of (a), the proof is similar, using the respective integral identities.

COROLLARY 2.8. It follows from Lemma 2.4 and Lemma 2.5, respectively, that bands of the first or the second kind, at x_0 are regular to the right of the band point in (a, b), i.e., in the differential equation (c) the function $w(x) \neq 0$ for $x > x_0 \in (a, b)$; hence their zeros (if they exist) separate each other in (x_0, b). If the differential equation (a) has an oscillatory solution in (a, b) and the hypotheses of Lemma 2.5 are satisfied, then every band of the first kind and every band of the second kind oscillates in (a, b).

THEOREM 2.10. Let $A(x) \leqq 0$ and suppose that $b(x)$ has the property (v) in (a, b). Then there exists at least one solution of the differential equation (a) $y(x) \neq 0$ in (a, b) having the following properties: y and y' are monotone functions of $x \in (a, b)$ and

$$\operatorname{sgn} y(x) = \operatorname{sgn} y''(x) \neq \operatorname{sgn} y'(x) \quad \text{for } x \in (a, b).$$

PROOF. The existence of a solution $y \neq 0$ for $y \in (a, b)$ satisfying the identity (1.13) follows from the proof of Theorem 1.13. Assume that this solution is positive, i.e. $y(x) > 0$ for $x \in (a, b)$. Then the identity (1.13) implies that $y''(x) > 0$ for $x \in (a, b)$. This means that y' is a monotone function of $x \in (a, b)$ and y has at most one extremum, namely a minimum. We show that this is not possible, i.e. that y is also a monotone function of $y \in (a, b)$. In fact, assume that $y'(\xi) > 0$, where $\xi \in (a,b)$. It follows from the proof of Theorem 1.13 that $y'(\xi) = \lim_{n \to \infty} y_n'(\xi)$. However, for $x_n > \xi$ we have $y_n'(\xi) < 0$, since $y_n''(\xi) > 0$ by the assumption. This is a contradiction, showing that at no point can $y'(x)$ be positive. It is easy to verify that $y'(x) = 0$ cannot hold at any point since $y''(x)>0$, and hence $y'(x)$ is an increasing function. Thus we have

$$\operatorname{sgn} y(x)=\operatorname{sgn} y''(x)\neq\operatorname{sgn} y'(x), \quad x\in(a, b). \tag{2.22}$$

The theorem is proved.□

REMARK 2.6. Let the interval $(a, b)\equiv(a, \infty)$. A solution $y(x)>0$ in (a, ∞) with the property (2.22) satisfies also $y'\to 0$ as $x\to\infty$.

Suppose this is false and let $y'(x)<k<0$. Integrating this inequality from x_0 to x, $x_0<x\in(a, b)$, leads to a contradiction for sufficiently great x.

THEOREM 2.11. Let $A(x)\leqq 0$, $A'(x)+b(x)\geqq 0$ and assume that $b(x)$ has the property (v) for $x\in(a, b)$. Then every solution y of the differential equation (a) with $y(x)\neq 0$, $y'(x)\neq 0$, $\operatorname{sgn} y(x)\neq \operatorname{sgn} y'(x)$ for $x\in(a, b)$ has $y''(x)$ monotone in (a, b). If $(a, b)\equiv(a, \infty)$, then $\operatorname{sgn} y(x)=\operatorname{sgn} y''(x)$ in (a, ∞) and, moreover, $\lim_{x\to\infty} y''(x)=0$.

PROOF. Let y be a solution of the differential equation (a) with, say, $y(x)>0$, $y'(x)<0$ for $x\in(a, b)$. Then it follows from the differential equation (a) that $y'''(x)=-2A(x)y'(x)-(A'(x)+b(x))y\leqq 0$ for $x\in(a, b)$. Therefore, $y(x)$ is a decreasing function of $x\in(a, b)$.

Suppose, moreover, that $(a, b)\equiv(a, \infty)$ and assume $y''(x_0)\leqq 0$. In view of $y''(x)$ being a decreasing function of $x\in(a, b)$, we have $y''(x)<0$ for $y>x_0$. Integrating the inequality $y''(x)<0$ from x_0 to x, $x_0<x$, we obtain $y'(x)<y'(x_0)$, $y(x)<y(x_0)+y'(x_0)\,(x-x_0)$. Hence, for sufficiently great x we get $y(x)<0$, a contradiction. Therefore $y''(x)>0$ for $x\in(a, b)$ and $\lim_{x\to\infty} y''(x)=0$. □

REMARK 2.7. If $(a, b)\equiv(a, \infty)$, the solution y with the property (2.22) in Theorems 2.10, 2.11 satisfies $\lim_{x\to\infty} y(x)=c\neq\pm\infty$.

THEOREM 2.12. Let $A(x)\leqq 0$, $A'(x)+b(x)\geqq 0$ and let $b(x)$ have the property (v) in (a, b). A necessary and sufficient condition for the differential equation (a) to be oscillatory in (a, b) is that

$$y(x)y'(x) \neq 0, \text{ sgn } y(x) \neq \text{sgn } y'(x), \quad x \in (a, b) \tag{2.23}$$

for all its non-trivial, non-oscillatory solutions y.

PROOF. Sufficiency is obvious since any solution of (a) having a zero must be oscillatory, otherwise it would violate the condition (2.23).

Necessity. Suppose that the differential equation (a) is oscillatory in (a, b) and let $y(x) \neq 0$ be a solution with no zeros in (a, b). Without loss of generality we may assume $y(x) > 0$ in (a, b). By Corollary 2.8, evidently $y(x)y'(x) \neq 0$ for $x \in (a, b)$. We have to show that $y'(x) < 0$ for $x \in (a, b)$. Now, Theorem 2.10 implies that there exists a solution $v(x) > 0$, $x \in (a, b)$, of the differential equation (a) satisfying (2.22), that is, $v'(x) < 0$ for $x \in (a, b)$. Assume the contrary, namely that $y'(x) > 0$ for $x \in (a, b)$. Let $x_0 \in (a, b)$ and let k be a positive constant such that $y(x_0) - kv(x_0) = 0$. Then the solution $u(x) = y(x) - kv(x)$ of the differential equation (a) vanishes at y_0, and is therefore oscillatory in (a, b). On the other hand, $u'(x) = y'(x) - kv'(x) > 0$ for $x \in (a, b)$, which yields a contradiction. Therefore $y'(x) < 0$ for $x \in (a, b)$ and the theorem is proved. □

REMARK 2.8. Lazer [86] and Vilari [142] were the first to derive a necessary and sufficient condition of the above type for the equation

$$y''' + py' + qy = 0\,, \tag{$\bar{\text{a}}$}$$

where p and q are continuous functions of $x \in (a, \infty)$. Their results were generalized by Gera [33]. We shall deal with this topic again in Chapter 2.

THEOREM 2.13. Let $A(x) \leqq 0$, $A'(x) + b(x) \geqq 0$ for $x \in (a, \infty)$ and let $b(x)$ be a function having the property (v) in (a, ∞). If the differential equation (a) has an oscillatory solution in (a, ∞), then every solution but one (up to linear dependence) is oscillatory and the non-oscillatory solution y has the following properties: $y(x) \neq 0$, $\text{sgn } y(x) = \text{sgn } y''(x) \neq \text{sgn } y'(x)$ for $x \in (a, \infty)$; y, y', y'' are monotone functions of $x \in (a, \infty)$, and $\lim_{x\to\infty} y'(x) = \lim_{x\to\infty} y''(x) = 0$.

PROOF. Theorem 2.12 implies that the differential equation (a) cannot have any solution without zeros in (a, b) and with the property $\operatorname{sgn} y(x_0) = \operatorname{sgn} y'(x_0)$ or $y'(x_0) = 0$, $x_0 \in (a, \infty)$. Every solution $y \neq 0$ in (a, ∞) has the property (2.23) and, by Theorem 2.11, also the property (2.22), and y, y', y'' are monotone functions of $x \in (a, \infty)$, while $\lim_{x\to\infty} y' = \lim_{x\to\infty} y'' = 0$. It remains to prove that there is at most one solution with no zeros in (a, ∞).

Assume the contrary. Let y and $\bar{y}$ be two solutions without zeros in (a, ∞). Let y_1, y_2 be some other solutions of the differential equation (a) satisfying, at $x_0 \in (a, \infty)$, the following conditions:

$$y_1(x_0) = y_1'(x_0) = 0,\ \ y_1''(x_0) \neq 0,\ \ y_2(x_0) = y_2''(x_0) = 0,$$
$$y_2'(x_0) \neq 0.$$

The solutions y_1, y_2, y evidently form a fundamental set of solutions of the differential equation (a), hence there exist constants c_1, c_2, c_3 such that $\bar{y} = c_1 y_1 + c_2 y_2 + c_3 y$. Since $\bar{y}$ is a solution with no zeros in (a, ∞), necessarily $\lim_{x\to\infty} \bar{y}'(x) = 0$. We show that this property is violated.

The solution $Y = c_1 y_1 + c_2 y_2$ is oscillatory in (a, ∞). Let $x_0 < x_1 < < x_2 < \ldots < x_i < \ldots$ be its zeros. The integral identity (1.6) implies the following equality at the points x_i, $i = 1, 2, \ldots$

$$Y'^2(x_i) = Y'^2(x_0) + 2 \int_{x_0}^{x} bY^2\, dt .$$

Therefore, $y(x)$ cannot be monotonically decreasing to zero, so the theorem is proved. □

LEMMA 2.6. Assume that $A(x) \leqq 0$, $A'(x) + b(x) \geqq 0$ for $x \in (a, \infty)$ and let $b(x)$ have the property (v) in (a, ∞). Then every non-trivial, non-oscillatory solution y of the differential equation (a) either has the property (2.22) in (a, ∞) and $y(x) \neq 0$ in (a, ∞), or there exists $\alpha > a$ such that $y(x)y'(x) \geqq 0$ for $x \geqq \alpha$ and $y(x) \neq 0$ for $x \geqq \alpha$.

PROOF. Let y be a non-trivial, non-oscillatory solution of the differential equation (a) in (a, ∞), and let, say, $y(x) > 0$, for $x \geqq \alpha$. Assume that $y'(x) > 0$ for $\alpha \leqq x < x_1$ and $y'(x_1) = 0$, also that y is a decreasing function of x for $x > x_1$. We show that this is impossible. From the differential equation (a) it would follow that $y'''(x) \leqq 0$ for $x > x_1$. At x_1, we have $y''(x_1) \leqq 0$ and sufficiently near to x_1, there exists a point $\bar{x} > x_1$ such that $y''(\bar{x}) < 0$. Integrate $y'''(x) \leqq 0$ twice between $\bar{x}$ and x, $x > \bar{x}$. This gives for y' the inequality $y'(x) \leqq y'(\bar{x}) + y''(\bar{x})\ (x - \bar{x})$, which implies that $y'(x) \to -\infty$ as $x \to \infty$. Thus the solution y would necessarily have a zero, which is a contradiction.

Now let $\alpha < x_1 < x_2$, $y'(x) > 0$ for $\alpha \leqq x < x_1$ and $y'(x_1) = y'(x_2) = 0$, while $y'(x) < 0$ for $x_1 < x < x_2$. Multiply the equation (a) by y' and integrate by parts from x_1 to x_2. After simple reductions we obtain

$$0 = -\int_{x_1}^{x_2} y''^2(t)\, dt + 2\int_{x_1}^{x_2} A(t) y'^2(t)\, dt +$$

$$+ \int_{x_1}^{x_2} [A'(t) + b(t)] y(t) y'(t)\, dt\, .$$

Thus we have a contradiction since the right-hand side is negative. Therefore, if y is a non-oscillatory solution in (α, ∞), y' cannot change its sign. The first part of the assertion of Lemma 2.6 is obvious when we take Theorems 2.10 and 2.11 also into consideration. □

THEOREM 2.14. Let $A(x) \leqq 0$, $A'(x) + b(x) \geqq 0$ for $x \in (a, \infty)$ and let $b(x)$ have the property (v) in (a, ∞).

If, moreover,

$$\int_{x_0}^{\infty} \left[A'(t) + b(t) - \frac{4}{3}\sqrt{\frac{2}{3}}\sqrt{-A^3(t)} \right] dt = +\infty\, , \qquad (2.24)$$

$a < x_0 < \infty$, then the differential equation (a) is oscillatory in (a, ∞).

PROOF. Let y be a non-trivial, non-oscillatory solution of the differential equation (a). According to Lemma 2.6, there exists α, $a < \alpha < \infty$, such that either

$$t(x) = \frac{y'(x)}{y(x)} \geqq 0 \quad \text{for } x \geqq \alpha\, , \qquad (2.25)$$

or

$$t(x) = \frac{y'(x)}{y(x)} < 0 \quad \text{for } x \in (a, \infty)\, . \qquad (2.26)$$

We shall show that the case (2.25) cannot arise.

Suppose, to the contrary, that (2.25) is true. It is easy to observe that the function $t(x)$ satisfies the following equation (the so-called Riccati second order non-linear equation)

$$t''(x)+3t'(x)t(x)=-(t^3(x)+2A(x)t(x)+A'(x)+b(x)).$$

Define the function $F(u, x)\overset{\mathrm{d}}{=}u^3+2A(x)u+A'(x)+b(x)$. Let us find the point u at which $F(u, x)$ attains a local minimum for every x. We get $u=\sqrt{(-2/3)A}$. At that point, clearly,

$$t''(x)+3t'(x)t(x)<-(A'+b)+\frac{2}{3\sqrt{3}}\sqrt{(-2A)^3}=$$

$$=-(A'+b)+\frac{4}{3}\sqrt{\frac{2}{3}}\sqrt{-A^3},$$

i.e.

$$\frac{\mathrm{d}}{\mathrm{d}x}\left(t'(x)+\frac{3}{2}t^2(x)\right)\leqq-(A'+b)+\frac{4}{3}\sqrt{\frac{2}{3}}\sqrt{-A^3}.$$

Hence, after integration, we get

$$t'(x)\leqq t'(a)+\frac{3}{2}t^2(a)-\frac{3}{2}t^2(x)-$$

$$-\int_a^x\left[A'+b-\frac{4}{3}\sqrt{\frac{2}{3}}\sqrt{-A^3}\right]\mathrm{d}t.$$

It follows from (2.24) that $t'(x)=y'(x)/y(x)$ would be negative for sufficiently great x, which contradicts the assumption. Therefore $y(x)$ is an oscillatory solution. The non-oscillatory solutions have the property (2.26), i.e. the property (2.22), and by Theorem 2.12, the differential equation (a) is oscillatory in (a, ∞). Theorem 2.13 implies that there is only one non-oscillatory solution up to linear dependence. Thus the theorem is proved. □

REMARK 2.9. A sufficient condition for the equation ($\bar{a}$) was derived by Ahmad and Lazer [1] under the assumptions $p(x)\leqq 0$ and $q(x)>0$, $a\leqq x<\infty$, while

$$\int_a^\infty \left[q(t) - \frac{2}{3\sqrt{3}}(-p(t))^{3/2}\right] dt = \varphi .$$

The method of proof is the same.

COROLLARY 2.9 Let the hypotheses of Theorem 2.14 be satisfied and, moreover, let $-k \leqq A(x) \leqq 0$ for $x \in (a, \infty)$, where $k > 0$ is a constant. Instead of (2.24), we may then assume

$$\int_{x_0}^\infty \left[b(t) - \frac{4}{3}\sqrt{\frac{2}{3}}\sqrt{-A^3(t)}\right] dt = \infty .$$

LEMMA 2.7. Suppose $A(x) \leqq 0$ and $b(x) - |A'(x)| \geqq k > 0$ for $x \in (a, \infty)$, k being a constant, and let $b(x)$ have the property (v) in (a, ∞). Let w_1 be a solution of the differential equation (b) with $w_1(x_0) = w_1'(x_0) = 0$, $w_1''(x_0) > 0$, $a < x_0 < \infty$.* Then $\lim_{x\to\infty} w_1(x) = \lim_{x\to\infty} w_1'(x) = \infty$ and $\lim_{x\to\infty} w_1''(x) > 0$.

PROOF. The integral identity (1.9) for w_1 becomes

$$w_1'' + 2Aw_1 - \int_{x_0}^x (A' + b)w_1 \, dt = w_1''(x_0) .$$

Substituting for w_1'' from the above identity in

$$(w_1 w_1')' = w_1 w_1'' + w_1'^2 ,$$

we obtain

$$(w_1 w_1')' = w_1 \left[w_1''(x_0) + \int_{x_0}^x (A'(t) + b(t))w_1(t) \, dt - \right.$$
$$\left. - 2A(x)w_1(x)\right] + w_1'^2(x) .$$

The last equality implies that $w_1(x) > 0$, $w_1'(x) > 0$ for $x > x_0$. From

* Sometimes we use the notation $\lim_{x\to\infty} f(x) = \infty$ as a convenient abbreviation for $f(x) \to \infty$ as $x \to \infty$.

(1.9) for w_1 it follows that $w_1''(x) \geqq w_1''(x_0) > 0$ for $x > x_0$. Integrating the last inequality twice from x_0 to x, $x > x_0$, we get $w_1'(x) \geqq w_1''(x_0)(x - x_0)$ and $w_1(x) \geqq \frac{1}{2} w_1''(x)(x - x_0)^2$, which implies that $w_1(x) \to \infty$, $w_1'(x) \to \infty$ as $x \to \infty$. From the differential equation (b) it follows that $w_1'''(x) \geqq 0$ for $x > x_0$ and hence $\lim_{x\to\infty} w_1''(x) > 0$ since $w_1''(x)$ is a non-decreasing function for $x > x_0$. The lemma is proved. □

REMARK 2.10. The same properties which w_1 possesses are also enjoyed by solutions w_2, w_3 of the differential equation (b) satisfying $w_2(x_0) = w_2''(x_0) = 0$, $w_2'(x_0) > 0$, $w_3'(x_0) = w_3''(x_0) = 0$, $w_3(x_0) > 0$. The proof is analogous.

THEOREM 2.15. Let the hypotheses of Lemma 2.7 hold true. If the differential equation (a) has an oscillatory solution in (a, ∞), then all the solutions of the differential equation (a) are oscillatory in (a, ∞) except for one solution y (up to linear dependence), which has the following properties: $y(x)y'(x)y''(x) \neq 0$, $\operatorname{sgn} y(x) = \operatorname{sgn} y''(x) \neq \operatorname{sgn} y'(x)$ for $x \in (a, \infty)$, y, y', y'' are monotone functions of $x \in (a, \infty)$ and $\lim_{x\to\infty} y(x) = \lim_{x\to\infty} y'(x) = \lim_{x\to\infty} y''(x) = 0$.

PROOF. From the properties of bands it follows that whenever a solution oscillates in (a, ∞), then all the solutions having at least one zero in (a, ∞) oscillate also.

Now let y_1, y_2, y_3, be a fundamental set of solutions of the differential equation (a) such that $y_1(x_0) = y_1'(x_0) = 0$, $y_1''(x_0) = 1$, $y_2(x_0) = y_2''(x_0) = 0$, $y_2'(x_0) = 1$, $y_3'(x_0) = y_3''(x_0) = 0$, $y_3(x_0) = 1$, $a < x_0 < \infty$.

The solutions y_1, y_2 are oscillatory in (a, ∞). The functions $w_1 = y_1y_2' - y_1'y_2 = w(y_1, y_2)$, $w_2 = y_1y_3' - y_1'y_3 = w(y_1, y_3)$, $w_3 = y_2y_3' - y_2'y_3 = w(y_2, y_3)$ are solutions of the differential equation (b) with the properties described in Lemma 2.7 and Remark 2.10. In fact, we have $w_1(x_0) = w_1'(x_0) = 0$, $w_1''(x_0) = -1$, $w_2(x_0) = w_2''(x_0) = 0$, $w_2'(x_0) = -1$, $w_3(x_0) = -1$, $w_3'(x_0) = 0$, $w_3''(x_0) = 2A(x_0)$. Therefore, by Lemma 2.7 and Remark 2.10, $w_1(x) < 0$, $w_2(x) < 0$, $w_3(x) < 0$ for $x > x_0$. Hence

$y_3(x)$ is also oscillatory in (x_0, ∞), because it belongs to a regular band with the solution y_2, which is oscillatory.

The general solution of the differential equation (a) is $y = c_1y_1 + c_2y_2 + c_3y_3$. We have $y(x_0) = c_3$, $y'(x_0) = c_2$, $y''(x_0) = c_1$. If at least one of the constants c_1, c_2, c_3 wanishes, then the solution y is oscillatory. Let $c_1c_2c_3 \neq 0$. Four cases can then arise:

a) $\operatorname{sgn} c_1 = \operatorname{sgn} c_2 = \operatorname{sgn} c_3$, *b*) $\operatorname{sgn} c_3 \neq \operatorname{sgn} c_2 = \operatorname{sgn} c_1$,
c) $\operatorname{sgn} c_3 = \operatorname{sgn} c_2 \neq \operatorname{sgn} c_1$, d) $\operatorname{sgn} c_1 = \operatorname{sgn} c_3 \neq \operatorname{sgn} c_2$.

It is evident that the solutions y in cases a), b) and c) are oscillatory in (a, ∞). In the cases a) and b) we have, in fact,

$$w(y_3, y) = c_2w(y_3, y_2) + c_1w(y_3, y_1) \neq 0 \quad \text{for} \quad x > x_0.$$

In the case c),

$$w(y, y_1) = c_3w(y_3, y_1) + c_2w(y_2, y_1) \neq 0 \quad \text{for} \quad x > x_0.$$

Therefore, solutions without zeros must be of the form d), that is, $\operatorname{sgn} y(x_0) = \operatorname{sgn} y''(x_0) \neq \operatorname{sgn} y'(x_0)$, while $y(x_0)y'(x_0)y''(x_0) \neq 0$. It follows that, for $x \in (a, \infty)$, $y(x)y'(x)y''(x) \neq 0$ and $\operatorname{sgn} y(x) = \operatorname{sgn} y''(x) \neq \operatorname{sgn} y'(x)$. Let us form the wronskian $W(y, y_1, y_2)$; obviously, $W(y, y_1, y_2) = -y(x_0)$. After expanding $W(y, y_1, y_2)$ by its first column we obtain

$$y''w(y_1, y_2) - y'w'(y_1, y_2) + y[w''(y_1, y_2) + 2Aw(y_1, y_2)] = -y(x_0)\,. \tag{2.27}$$

According to Lemma 2.7 we have $w(y_1, y_2) \to -\infty$, $w'(y_1, y_2) \to -\infty$ as $x \to \infty$. We now show that $w''(y_1, y_2) + 2Aw(y_1, y_2) \to -\infty$ as $x \to \infty$. Observe that $w(y_1, y_2) = w_1$. The assertion follows from (1.9) for w_1 and from the assumption that $b - |A'| \geqq k > 0$ for $x \in (a, \infty)$. Taking this into account, we deduce from (2.27) that $\lim_{x \to \infty} y(x) = 0$.

Uniqueness of the solution without zeros follows from Theorem 2.13. Thus the theorem is proved. □

REMARK 2.11. The first part of the proof of Theorem 2.15 could be shortened in view of the proof of Theorem 2.13. The proof follows a different method, however, and therefore it has been given in full.

6. *Sufficient Conditions for Oscillatoricity of Solutions of the Differential Equation (a) in the Case $A(x)\geqq 0$, $x\in(a,\infty)$*

The results presented in this section were published simultaneously in 1966 in papers by Lazer [86] and Greguš [52]. Consequently some of them overlap.

LEMMA 2.8. Let $A(x)\geqq 0$, $A'(x)+b(x)\geqq 0$ and let $b(x)$ have the property (v) in (a,∞). If y is a non-trivial, non-oscillatory solution of the differential equation (a) in (a,∞) with $y(x)\geqq 0$ for $x>\alpha$, $a<\alpha<\infty$ and if, also

$$y(c)y''(c)-\frac{1}{2}y'^2(c)+A(c)y^2(c)=F(y(c))\leqq 0,$$

$$\alpha\leqq c<\infty,$$

then there exists a number $d\geqq c\in(a,\infty)$ such that $y(x)>0$, $y'(x)>0$, $y''(x)>0$ and $y'''(x)\leqq 0$ for $x\geqq d\in(a,\infty)$.

PROOF. It follows from the integral identity (1.6) for the solution y that the function

$$F(y(x))=F(y(c))-\int_c^x by^2\,dt$$

is non-positive, decreasing for $x\geqq c$ and vanishes at those points where both y' and y vanish. If $y\geqq 0$ for $x>c$ is a non-oscillatory solution of the differential equation (a), then there exists $c_1\geqq c$ such that $y(x)>0$ for $x\geqq c_1$. Otherwise y would necessarily have more than one double zero, which is evidently impossible in view of the properties of bands.

If $y'(\bar{c}_1)=0$, $c_1\leqq\bar{c}_1<\infty$, then

$$F(y(\bar{c}_1))=y(\bar{c}_1)\bar{y}''(\bar{c}_1)+A(\bar{c}_1)y^2(\bar{c}_1)<0\,,$$

and hence $y''(\bar{c}_1)<0$. Thus $y'(x)$ cannot have more than one zero in the interval $\bar{c}_1 \leqq x < \infty$. There exists therefore a point $c_2 < c_1$ such that $y(x) > 0$, $y'(x) \neq 0$ for $x \geqq c_2$. We now show that $y'(x) > 0$ for $x > c_2$. Assume the contrary, i.e. that $y'(x) < 0$ for $x \geqq c_2$. Three cases can arise:

a) Let $y''(x) \leqq 0$ for $x \geqq \bar{c}_2 \geqq c_2$, then $y'(x) \leqq y'(\bar{c}_2) < 0$ for $x \geqq \bar{c}_2$, so that y could become negative for sufficiently great x, which is a contradiction.

b) Let $y''(x) \geqq 0$ for $x \geqq \bar{c}_2 \geqq c_2$. Since $y'(x) < 0$ for $x \geqq \bar{c}_2$, we have $\lim_{x \to \infty} y'(x) = 0$ and hence

$$\lim_{x \to \infty} F(y(x)) = \lim_{x \to \infty} \left(yy'' - \frac{1}{2} y'^2 + Ay^2 \right) =$$

$$= \lim_{x \to \infty} (yy'' + Ay^2) \geqq 0 ,$$

contradicting the assumption that $F(y(x))$ is a non-positive and decreasing function of $x \geqq c$.

c) Let $y''(x)$ change its sign for sufficiently great x. Then, given any $\varepsilon > 0$, there exist sufficiently great numbers x for which $0 > y'(x) > -\varepsilon$, and therefore there are points $\bar{x}$ at which $y'(x)$ attains local maxima with $y''(\bar{x}) = 0$. Hence, for arbitrarily great $\bar{x}$, the following inequality is true:

$$-F(y(\bar{x})) \leqq \varepsilon^2 - Ay^2 \leqq \varepsilon^2 .$$

As a consequence,

$$\lim_{x \to \infty} F(y(x)) \geqq 0 .$$

This yields again a contradiction, hence $y'(x) > 0$ for $x \geqq c_2$.

It follows then from the differential equation (a) that $y'''(x) \leqq 0$ for $x \geqq c_2$, hence $y''(x)$ does not change its sign for $x \geqq c_2$. If we had $y''(x) < 0$ for $x \geqq d \geqq c_2$, we would also have

$$y''(x) \leqq y''(d) < 0 \quad \text{for} \quad x \geqq d ,$$

so that $\lim_{x \to \infty} y'(x) = -\infty$, a contradiction. Thus, necessarily, $y''(x) > 0$ for $x \geqq d$, so the lemma is proved. □

LEMMA 2.9. If y is a solution of the differential equation (a) with $y(x)>0$, $y'(x)>0$, $y'''(x)\leqq 0$ for $x\geqq\alpha>a$, then

$$\liminf_{x\to\infty}\frac{y(x)}{xy'(x)}\geqq\frac{1}{2}\,.$$

PROOF. Define

$$G(x)\overset{\mathrm{d}}{=}(x-\alpha)y(x)-\frac{(x-\alpha)^2}{2}\,y(x)\,.$$

We have

$$G(\alpha)=0,\ \ G'(x)=y(x)-\frac{(x-\alpha)^2}{2}\,y''(x)\,.$$

By Taylor's theorem,

$$y(x)=y(\alpha)+y'(\alpha)(x-\alpha)+\frac{(x-\alpha)^2}{2}\,y''(\xi),\ \ \alpha<\xi<x\,.$$

Hence

$$G'(x)=y(\alpha)+y'(\alpha)(x-\alpha)+\frac{(x-\alpha)^2}{2}\,[y''(\xi)-y''(x)]\,.$$

Since $y'''(x)\leqq 0$ for $x\geqq\alpha$, we have $y''(\xi)\geqq y''(x)$, and hence $G'(x)>0$ for $x>\alpha$. But since $G(a)=0$, $G'(x)>0$, we have $G(x)>0$ and the following inequality:

$$\frac{y(x)}{(x-\alpha)y'(x)}>\frac{1}{2}\,,\qquad x>\alpha\,.$$

The assertion of the lemma follows from the last inequality. □

REMARK 2.12. Lemma 2.9 is true irrespective of whether y with the assumed properties is a solution of the differential equation (a) or not. It suffices that y satisfies the given assumptions and that $y'''(x)$ is a continuous function of $x\in\langle\alpha,\infty)$.

THEOREM 2.16. Let the coefficients of the differential equation (a) satisfy the hypotheses of Lemma 2.8 in (α,∞) and let, for some number $m<\frac{1}{2}$, the second order differential equation

$$v'' + [2A(x) + m(A'(x) + b(x)]\ v = 0$$

be oscillatory in (a, ∞). Then the differential equation (a) is also oscillatory in (a, ∞) and, moreover, every solution of the differential equation (a) with the property $F(y(c)) \leqq 0$, $a < c < \infty$, is oscillatory in (a, ∞).

PROOF. Suppose that a non-trivial solution y of the differential equation (a) satisfies $F(y(c)) \leqq 0$ and is non-oscillatory in (a, ∞). Let, for example, $y(x) \geqq 0$ for $x \geqq c$. By Lemma 2.8, there exists $d \geqq c$ such that

$$y(x) > 0,\ y'(x) > 0,\ y''(x) > 0,\ y'''(x) \leqq 0 \text{ for } x \geqq d\,. \tag{2.28}$$

According to Lemma 2.9 we have

$$\liminf_{x \to \infty} \frac{y(x)}{xy'(x)} \geqq \frac{1}{2}\,.$$

Let $m < \frac{1}{2}$. Then there exists $d_1 \geqq d$ such that $y(x)/y'(x) > mx$ for $x \geqq d_1$. Rewrite the differential equation (a) as a system:

$$v'' + 2Av + (A' + b)y = 0, \quad y' = v\,. \tag{2.29}$$

The first differential equation of this system (2.29) can be written in the form

$$v'' + \left[2A(x) + (A'(x) + b(x))\,\frac{y(x)}{v(x)}\right] v(x) = 0\,.$$

Evidently,

$$2A(x) + [A'(x) + b(x)]\,\frac{y(x)}{v(x)} = 2A(x) +$$

$$+ [A'(x) + b(x)]\,\frac{y(x)}{y'(x)} > 2A(x) + m(A'(x) + b(x))$$

for $x \geqq d$.

In view of the oscillatoricity of solutions of the differential equation

$$u'' + [2A + m(A' + b)]\ u = 0$$

it follows from the Sturm theorem that the solutions of the second order differential equation

$$v'' + \left[2A + (A' + b)\frac{y}{v}\right] v = 0 \tag{2.30}$$

are oscillatory. This contradicts (2.28), where $y'(x) = v(x)$. Theorem 2.16 is thus proved. □

THEOREM 2.17. **Suppose $b(x)$ has the property (v) in (a, ∞) and let the differential equation (a) have at least one oscillatory solution in (a, ∞). A necessary and sufficient condition for a non-trivial solution y to be non-oscillatory in (a, ∞) is that $F(y(x)) > 0$ for $x \geqq \alpha > a$.**

PROOF. Sufficiency follows immediately from the integral identity (1.6).

Necessity. It is sufficient to show that if y is a non-trivial solution of the differential equation (a) with the property $F(y(c)) \leqq 0$, $a < c < \infty$, then it is oscillatory in (a, ∞). If $y(c) = 0$, the assertion follows from the properties of bands. Let $y(c) \neq 0$ and let $\bar{y}$ be a solution of the differential equation (a) with $\bar{y}(c) = 0$, $\bar{y}'(c) = y(c)$, $\bar{y}''(c) = y'(c)$. The solution $\bar{y}(x)$ is evidently oscillatory in (a, ∞), because it belongs to the band at c. Consequently, for arbitrary constants $c_1 \neq 0$, $c_2 \neq 0$ we have

$$\begin{aligned} F(c_1 y(c) + c_2 \bar{y}(c)) &= c_1^2 F(y(c)) + c_1 c_2 y(c)\bar{y}(c) + \\ &+ y(c)\bar{y}(c) - y'(c)\bar{y}'(c) + 2A(c)y(c)\bar{y}(c) + c_2^2 F(\bar{y}(c)) = \\ &= c_1^2 F(y(c)) + c_1 c_2 [y(c)y'(c) - y(c)y'(c)] - \frac{c_2^2}{2} y^2(c) = \\ &= c_1^2 F(y(c)) - \frac{1}{2} c_2^2 y^2(c) \leqq 0 . \end{aligned} \tag{2.31}$$

The wronskian of the solutions y, $\bar{y}$ is $W(y, \bar{y}) = y\bar{y}' - \bar{y}y'$. If $W(y, \bar{y})\ (d) = 0$, $d > c$, then there are constants c_1, c_2 such that

$$c_1 y(d) + c_2 \bar{y}(d) = 0,$$

$$c_1 y'(d) + c_2 \bar{y}'(d) = 0 ,$$

where $c_1^2+c_2^2\neq 0$. Writing $z\overset{d}{=}c_1y+c_2\bar{y}$ we have $F(z(d))=0$, but $F(z(c))=c_1^2F(y(c))-1/2c_2^2y^2(c)\leqq 0$ by (2.31). From the integral identity (1.6) it follows that

$$0=F(z(d))=F(z(c))-\int_c^d bz^2\,\mathrm{d}t<0\,.$$

Thus $W(y,\bar{y})\neq 0$ for $x>c$, therefore y is an oscillatory solution of the differential equation (a), so the theorem is proved. □

REMARK 2.13. The integral identity (1.13), in which we may put $b=\infty$, shows that there exists at least one solution of the differential equation (a) with no zeros in (a,∞), satisfying

$$F(y)=yy''-\frac{1}{2}y'^2+Ay^2>0 \text{ for } x>\alpha,\ a<\alpha<\infty\,.$$

7. *Conjugate Points, Principal Solutions and the Relationship between the Adjoint Differential Equations (a) and (b)*

Let y be solution of the differential equation (a) with $y(\alpha)=0$, $a<\alpha<b$, and suppose it has at least $n+2$ zeros in $\langle\alpha,b)$. Denote these zeros by $\alpha_1,\alpha_2,\dots,\alpha_{n+2}$ $(\alpha=\alpha_1\leqq\alpha_2\leqq\alpha_3\leqq\dots\leqq\alpha_{n+2})$; by the n-th conjugate point of the point α we shall mean the least possible value of α_{n+2}, taken over all the solutions y of the differential equation (a) with $y(\alpha)=0$. This definition of conjugate points is justified by the following reasoning.

First of all, suppose that there exists a solution of the differential equation (a) with $y(\alpha)=0$, having at least $n+2$ zeros in $\langle a,b)$. If y_1, y_2 are two linearly independent non-trivial solutions of (a) with $y_1(\alpha)=0$, $y_2(\alpha)=0$, then all the solutions y with $y(\alpha)=0$ can be written in the form $y=Ay_1+By_2$. If only finitely many of these have $n+2$ zeros, the existence of the n-th conjugate point does not need any proof. If there are infinitely many such solutions, choose an infinite sequence $\{y_\nu(x)\}$ of them, where $y_\nu=A_\nu y_1+B_\nu y_2$. When we normalize these functions y_ν by $A_\nu^2+B_\nu^2=1$, the set $\{y_\nu\}$ evidently becomes locally uniformly bounded and equicontinuous. There exists therefore a sequence converging to some solution $y_0(x)$ of the differential equation (a). Let β be the infimum of the points α_{n+2} for such a subsequence of solutions y_ν. Since every limit point of zeros of solutions $y_\nu(x)$ belonging to the convergent sequence is a limit point of the solution $y_0(x)$, we have $y_0(\beta)=0$. If $\beta=\alpha$, then $y_0(x)\equiv 0$, otherwise $y_0(x)$ would have a triple zero α. However, this is impossible since $A_\nu^2+B_\nu^2=1$, hence $\beta>\alpha$, which proves the existence of the n-th conjugate point.

The solution $y_0(x)$ determining the n-th conjugate point is called the *principal solution* of the differential equation (a).

We are going to define two classes of differential equations of the form (a) and to study the properties of their solutions, using the properties of principal solutions.

Equations of class I. A differential equation of the form (a) belongs to class I if every solution y with $y(\alpha)=y'(\alpha)=0$, $y''(\alpha)>0$, $a<\alpha<b$, satisfies also $y(x)>0$ for $x\in(a, \alpha)$.

Equations of class II. A differential equation of the form (a) belongs to class II if every solution y with $y(\alpha)=y'(\alpha)=0$, $y''(\alpha)>0$, $a<\alpha<b$, satisfies also $y(x)>0$ for $x>\alpha$.

REMARK 2.14. The notion of conjugate points was introduced by Hanan [61], who in the same paper derived some criteria for oscillatoricity. The notion was defined for a differential equation of the form

$$y'''+p(x)y''+q(x)y'+r(x)y=0\,,$$

where p, q, r are continuous functions in $(0, \infty)$.

We are going now to give a series of criteria which make it possible to decide whether a given differential equation of the form (a) belongs to class I or class II.

First, it follows from § 1 of Section 5 that whenever the function $b(x)$ has the property (v) in (a, b), then the differential equation (a) is in class I and the differential equation (b) is in class II in (a, b).

LEMMA 2.10. Let the differential equation (a) belong to class I (II) in (a, b). Let $y_0(x)$ $(z_0(x))$ be the principal solution of the differential equation (a) ((b)) and let $\eta_n(x)$ be the n-th conjugate point to

$\alpha \in (a, b)$. Then the number of all zeros, counted according to multiplicity, of the solution $y_0(x)$ $(z_0(x))$ at α and $\eta_n(\alpha)$ is exactly three.

PROOF. Suppose that $y_0(x)$ has simple zeros both at α and at $\eta_n(\alpha)$ and assume, with no loss of generality, that $y_0(x) > 0$ for $x \in (\beta, \eta_n(\alpha))$, where β is the first zero of the solution $y_0(x)$ to the left of $\eta_n(\alpha)$. Let $v(x)$ be a solution of the differential equation (a) with $v(\alpha) = 0$, $v(\eta_0(\alpha)) \neq 0$. The existence of such a solution $v(x)$ follows from the properties of bands and from the relation between solutions of mutually adjoint differential equations. Construct the function $w(x) = y_0(x) + \varepsilon v(x)$ with sufficiently small ε. Then $w(\alpha) = 0$ and other zeros of the solution $w(x)$ are sufficiently near to zeros of $y_0(x)$. If we choose ε so that $\varepsilon v(\eta_n(\alpha))$ be negative, then $w(\eta_n) = v(\eta_n)$ is also negative and $y_0(x) > 0$ in (β, η_n). Thus the $(n+2)$-nd zero of $w(x)$ lies to the left of the $(n+2)$-nd zero of $y_0(x)$. However, this is not possible since the solution $y_0(x)$ is principal. Therefore, y has a double zero at α or at $\eta_n(\alpha)$ (or at both). In wiew of the differential equation (a) being in class I, $y_0(x)$ cannot have a double zero at $\eta_n(\alpha)$. The lemma is thus proved for the case in which the differential equation (a) belongs to class I. In the other case, the proof is similar. □

REMARK 2.15. We shall prove in Chapter II, § 1 that if the second order differential equation

$$y'' + 2A(x)y = 0$$

is disconjugate in (a, b), i.e. if every one of its solutions has at most one zero in (a, b), and if $A'(x) + b(x) \geqq 0$ $(A'(x) - b(x) \leqq 0)$ for $x \in (a, b)$, then the differential equation (a) is in class I (class II). Moreover, under the stated assumptions, the following is true:

If y is a solution of the differential equation (a) with $y'(\alpha) = 0$, $y(\alpha) \geqq 0$, $y''(\alpha) > 0$, then $y'(x) < 0$ for $x \in (a, \alpha)$.

REMARK 2.16. It follows from Lemma 2.10 that n conjugate points to a given $\alpha \in (a, b)$, for a differential equation of class I, are zeros of a solution $y_1(x)$ of the differential equation (a), having a double zero at α. If this solution is normalized by $y_1''(\alpha) = 1$, it is usually referred to as the *first principal solution* of the differential equation (a) and denoted by $y(x, \alpha)$. The first n conjugate points (if they exist) are zeros of the solution $y(x, \alpha)$.

Analogously, Lemma 2.10 implies that n conjugate points, for a differential equation of class II, are points at which n uniquely determined (up to linear dependence) solutions $y_v(x)$ of the given differential equation with $y_v(\alpha)=0$ have a double zero, $v=1, 2, \ldots, n$.

The above results are collected in the following two theorems.

THEOREM 2.18. If the differential equation (a) is in class I in (a, b) and if y is its solution with $y(\alpha)=0$, $a<\alpha<b$, having at least $n+2$ zeros in $\langle\alpha, b)$, then the first principal solution $y(x, \alpha)$ has n + 2 zeros at $\alpha<\eta_1<\eta_2<\ldots<\eta_n$ (including multiplicity at $x=\alpha$) and every solution vanishing at α has fewer than $v+2$ zeros in the interval $\langle\alpha, \eta_v(\alpha)\rangle$.

THEOREM 2.19. If the differential equation (a) is in class II in (a, b) and if y is its solution satisfying $y(\alpha)=0$, $a<\alpha<b$, and having at least $n+2$ zeros in $\langle\alpha, b)$, then there are n points $\eta_1<\eta_2<\ldots<\eta_n\in(a, b)$ and n uniquely determined solutions $y_v(x)$ of the differential equation (a) such that

a) $y_v(x)$ has a simple zero at $x=\alpha$ and a double zero at $x=\eta_v(\alpha)$,

b) $y_v(x)$ has exactly $v+2$ zeros in $\langle\alpha, \eta_v(\alpha)\rangle$ (including multiplicity),

c) every other solution vanishing at α has fewer than $v+2$ zeros in $\langle\alpha, \eta_v(\alpha)\rangle$, $v=1, 2, \ldots, n$.

The relationship between conjugate points of adjoint differential equations (a) and (b) is implied by the following lemma.

LEMMA 2.11. The differential equation (a) is in class I if and only if its adjoint equation (b) is in class II.

The proof of the lemma follows immediately from Theorem 1.3. The following theorem is a consequence of Lemma 2.11 and Theorem 2.19.

THEOREM 2.20. If the differential equation (a) is in class I, its conjugate points are the same as the conjugate points of the adjoint differential equation (b).

REMARK 2.17. If the differential equation (a) is in class I, then it is oscillatory if and only if there exist infinitely many points conjugate to α, $a<\alpha<b$ whose limit point is b.

COROLLARY 2.10. If the differential equation (a) is in class I, then the set of solutions $y=c_1y_1+c_2y_2$ with $y_1(\alpha)=y_1'(\alpha)=0$, $y_1''(\alpha)\neq 0$, $y_2(\alpha)=y_2''(\alpha)=0$, $y_2'(\alpha)\neq 0$, $a<\alpha<\infty$, where c_1, c_2 are constants not both zero, is a band satisfying the equation (c), where $w=w(x)=y_1y_2'-y_1'y_2$ is a solution of the differential equation (b) having a double zero at α and $w(x)\neq 0$ for $x>\alpha$.

Corollary 2.10 enables us to apply the whole theory of bands to differential equations of the form (a) belonging to class I. For example, if one solution of (a) oscillates in (a, b), then so do all the solutions having a zero, etc.

A direct consequence of the theory of bands is the following theorem.

THEOREM 2.21. Let the differential equation (a) be in class I and let y be its solution with $y(\alpha)=0$, $y(x)\neq 0$ for $x>\alpha$, $a<\alpha<b$. Then every solution of (a) has at most two (including multiplicity) zeros in (α, b), i.e., the differential equation (a) is disconjugate in (α, b).

COROLLARY 2.11. If the differential equation (a) is in class I in (a, b), then between any two neighbouring zeros of an arbitrary solution of (a) there are at most two zeros of any other solution.

The assertion follows from Theorem 2.21 and from the properties of bands.

We show now that the zeros of two principal solutions, namely those of the first two principal solutions, separate each other in a certain sense.

THEOREM 2.22. Let differential equation (a) be in class I in (a, b) and let $u(x)=y(x, \alpha)$ and $v(x)=y(x, \beta)$ be two principal solutions of (a) $\alpha<\beta\in(a, b)$. Further, let $\eta_n(\beta)$ be n-th conjugate points of the solutions $u(x)$ and $v(x)$. If $v(x)$ has exactly one zero in the interval $(\eta_k(\alpha), \eta_{k+1}(\alpha))$ and $u(x)$ has exactly one zero in $(\eta_k(\beta), \eta_{k+1}(\beta))$, then the zeros of the solutions $u(x)$ and $v(x)$ either separate each other one by one, or between every two zeros of $u(x)$ there are exactly two zeros of $v(x)$ in $(\eta_k(\alpha), b)$; or, respectively, the zeros of $u(x)$ and $v(x)$ either separate each other one by one in $(\eta_k(\beta), b)$, or between every two zeros of $v(x)$ there are exactly two zeros of $u(x)$ in $(\eta_k(\beta), b)$.

PROOF. To be specific, let $v(x)$ have exactly one zero in some interval $(\eta_k(\alpha), \eta_{k+1}(\alpha))$. Suppose then that there exists $j>k$ such that $v(x)$ has no zero in the interval $(\eta_j(\alpha), \eta_{j+1}(\alpha))$. By Lemma 1.2, there exists a function $w(x)=u(x)-\lambda v(x)$ having a double zero in that interval. Then $w(\eta_k(\alpha))=-\lambda v(\eta_k(\alpha))$ and $w(\eta_{k+1}(\alpha))=-\lambda v(\eta_{k+1}(\alpha))$. Since $v(x)$ has exactly one zero in $(\eta_k(\alpha), \eta_{k+1}(\alpha))$, we deduce that $v(\eta_k(\alpha))$ and $v(\eta_{k+1}(\alpha))$ have opposite signs, which means that $w(x)$ has another zero to the left of its double zero. This contradicts the assumption that the differential equation (a) belongs to class I in (a, b).

In the second case, assuming that $u(x)$ has no zero in $(\eta_j(\beta), \eta_{j+1}(\beta))$, $j>k$, we again get a contradiction.

If the zeros of the solutions u and v do not separate each other one by one, they do so in pairs. The theorem is thus proved. □

Theorem 2.22 implies the following corollary.

COROLLARY 2.12. Let the differential equation (a) be in class I in (a, b) and let $\eta_n(\alpha)$ and $\eta_n(\beta)$ be conjugate points to α and β, respectively. If $\eta_k(\alpha)<\beta<\eta_{k+1}(\alpha)$ for some positive integer $k(\eta_0(\alpha)=\alpha)$, then $\eta_n(\beta)\leqq\eta_{k+n+2}(\alpha)$.

Together with the differential equation (a), consider also the differential equation (a_2).

Using Corollary 2.12, we shall prove the following comparison theorem.

THEOREM 2.23. Let the differential equation (a) be in class I in (a, b) and assume that

$$b_1(x) \geqq b(x)\,, \quad a < x < b\,,$$

while $b_1(x) \not\equiv b(x)$ in any subinterval of (a, b).

If $\eta_n(\alpha)$ and $\eta_n'(\alpha)$, $a < \alpha < b$, are n-th conjugate points of (a) and (a_2) to α in (a, b), then the differential equation (a_2) belongs to class I and

$$\eta_n'(\alpha) < \eta_{2n-1}(\alpha)\,. \tag{2.32}$$

PROOF. To prove the first part of the theorem, suppose the contrary, i.e. that the differential equation (a_2) does not belong to class I. Let v be a solution of (a_2) such that $v(\beta) = v'(\beta) = 0$, $v''(\beta) > 0$, $a < \alpha < \beta$, and let α be the first zero of v to the left of β.

Let u be a solution of the differential equation (b) adjoint to (a) with $u(\alpha) = u'(\alpha) = 0$, $u''(\alpha) > 0$. The differential equation (b) is in class II, thus $u(\alpha) > 0$ for $x > \alpha$. Multiply the equation (b) by the function $v(x)$ and the equation (a_2) by $u(x)$, subtract and integrate from α to β; we thus obtain

$$[vu'' - v'u' + v''u + 2Avu]_\alpha^\beta = -\int_\alpha^\beta (b_1 - b)uv \, dt\,,$$

i.e.

$$v''(\beta)u(\beta) = -\int_\alpha^\beta (b_1 - b)\, uv \, dt\,.$$

Evidently, the left-hand side is positive and the right-hand side is negative, which is a contradiction, proving that the differential equation (a_2) belongs to class I.

Let $u(x) = y(x, \alpha)$ be the first principal solution of the differential equation (a) and let $\beta = \eta_1(\alpha)$. By Theorem 2.20, there exists a principal solution $w(x)$ of the differential equation (b) such that $w(\alpha) = w(\beta) = w'(\beta) = 0$ and $w(x) \neq 0$ for $\alpha < x < \beta$. Let $v(x) = z(x, \alpha)$ be the first principal solution of the differential equation (a_2). We begin by showing that v has a zero in (α, β). Suppose that $v(x) > 0$ in (α, β). If we multiply the equation (b) by $v(x)$ and (a_2) by $w(x)$, subtract and integrate from α to β, we get

$$v(\beta)w''(\beta) = -\int_\alpha^\beta (b_1 - b)vw \, dt\,. \tag{2.33}$$

Since $w''(\beta)>0$, the left-hand side of (2.33) is positive or zero if $v(\beta)=0$; the right-hand side, however, is negative unless $b_1\equiv b$ identically in (α, β). This is a contradiction, therefore $v(x)$ must have a zero in (α, β), that is,

$$\eta_1'(\alpha)<\eta_1(\alpha)\,. \tag{2.34}$$

We shall now prove (2.32) by induction. Let $\beta_1=\eta_1'(\alpha)$ and let $u_1(x)=y(x, \beta_1)$ and $v_1(x)=z(x_1, \beta_1)$ be the first principal solutions of (a) and (a_2), respectively. The solutions $v(x)$ and $v_1(x)$ belong to the same band of the differential equation (a_2) which is in class I, hence their zeros to the right of the band point separate each other. Since $v_1(x)$ has a double zero at β_1 and $v(x)$ has a simple zero at the same point, according to Theorem 1.8 and Corollary 2.11, the first zero of $v(x)$ to the right of β_1 is nearer to β_1 than the first zero of $v_1(x)$ to the right of β_1. Hence we have

$$\eta_{n+1}'(\alpha)<\eta_n'(\beta_1)<\eta_{n+2}'(\alpha)\,. \tag{2.35}$$

From Corollary 2.12 it follows that

$$\eta_n(\beta_1)<\eta_{n+2}(\alpha)\,. \tag{2.36}$$

Now assume that the inequality (2.32) holds for $n=k$. We prove its validity for $n=k+1$. The right-hand inequality of (2.35) shows that

$$\eta_{k+1}'(\alpha)<\eta_k'(\beta_1)\,,$$

and (2.32) for $n=k$ implies that

$$\eta_k'(\beta_1)<\eta_{2k-1}(\beta_1)\,.$$

It also follows from the inequality (2.36) for $n=k$ that

$$\eta_{2k-1}(\beta_1)<\eta_{2k+1}(\alpha)\,.$$

Combining the last three inequalities gives

$$\eta_{k+1}'(\alpha)<\eta_{2k+1}(\alpha)\,,$$

which establishes (2.32) for $n=k+1$. As we have already proved (2.32) for $n=1$, it holds for every n. The theorem is thus proved. □

It follows from Theorem 2.23 that if the differential equation (a) has infinitely many conjugate points, then the differential equation (a_2) has the same property. In view of properties of bands, we have the following theorem.

THEOREM 2.24. Let the differential equation (a) be in class I in (a, b) and assume that $b_1(x) \geqq b(x)$, while $b_1(x) \not\equiv b(x)$ in any subinterval of (a, b). If the differential equation (a) is oscillatory in (a, b), then so is the differential equation (a_2). □

If the differential equation (a) is in class I, its adjoint differential equation (b) is in class II. In view of this fact, the results about the differential equation (a) can be transferred to the differential equation (b) by a suitable transformation, taking into account the relations between conjugate points of both mutually adjoint differential equations, consequently we can state the following results concerning the differential equation (b), in class II.

THEOREM 2.25. Let the differential equation (b) belong to class II. Let $a < \alpha < b$ be a band point of the differential equation (b). Then the zeros of the band to the left of α separate each other in (a, α). □

THEOREM 2.26. If the differential equation (b) is in class II in (a, b), then between any two consecutive zeros of an arbitrary solution of (b) there are at most two zeros of any other solution of (b). □

THEOREM 2.27. If the differential equation (b) is in class II in (a, b) and if u and v are such non-trivial independent solutions of (b) that $u(\alpha) = v(\beta) = 0$, $a < \alpha < \beta < b$, then the number of zeros in (a, α) of the solution u differs from that of v by at most two. □

THEOREM 2.28. Let the differential equation (b) be in class II in (a, b) and let u be a solution of (b) with $u(\alpha) = 0$, $a < \alpha < b$, having $n+2$ zeros in $\langle \alpha, \beta \rangle$, $\alpha < \beta < b$. If $u(\beta) = u'(\beta) = 0$, then β is the n-th conjugate point to α for the differential equation (b).

PROOF. Suppose that the assertion of Theorem 2.28 is false, i.e., assume $\eta_n(\alpha) < \beta$. Then in view of Theorem 2.19, there exists a solution v of the differential equation (b) with $v(\alpha) = v(\eta_n) =$

$v'(\eta_n)=0$. Since $u'(\alpha)$ and $v'(\alpha)$ are non-zero, there is a constant m such that the function $w=u-mv$ has a double zero at α. Without loss of generality we may assume that $u'(\alpha)>0$, $v'(\alpha)>0$. Then $m>0$ also. Since (b) is in class II, we have $w(x)\neq 0$ for $x>\alpha$, hence $u(x)$ and $mv(x)$ do not intersect for $x>\alpha$. By Theorem 2.26, between any two neighbouring zeros of u there are at most two zeros of mv, and the solutions u and mv do not intersect. Therefore, the solution $v(x)$ cannot have more than $n+1$ zeros in the interval $\langle\alpha, \beta\rangle$, and hence $v(x)$ cannot be a principal solution of the differential equation (b). Thus, $\beta=\eta_n(\alpha)$. □

THEOREM 2.29. If the differential equation (b) is in class II in (a, b), if z is a solution of the differential equation (b) with $z(\alpha)=0$, $a<\alpha<b$, and if z has at least $n+2$ zeros in (α, b), then n points conjugate to α in $\langle\alpha, b)$ separate (one by one) the zeros of the solution z.

PROOF. Suppose that there exists a solution of the differential equation (b) with $z(\alpha_1)=0$ and that its further zeros are $\alpha_1=\alpha<\alpha_2<\ldots<\alpha_{n+1}<\alpha_{n+2}\in(a, b)$. Let v be a solution of (b) with $v(\alpha_1)=v'(\alpha_1)=0$, $v''(\alpha_1)>0$, that is, $v(x)>0$ for $x>\alpha_1$. If α_{n+2} is not a double zero of the solution z, then there exists, by Lemma 1.2, a constant λ such that the solution $w(x)=z(x)-\lambda v(x)$ has a double zero at some point ξ, $\alpha_{n+1}<\xi<\alpha_{n+2}$. Without loss of generality one can assume that $z(x)>0$ in $(\alpha_{n+1}, \alpha_{n+2})$; then λ is also positive. It is easy to see that the function $u(x)=\lambda v(x)$ intersects every positive segment of the solution z twice. If this were not the case in some interval where $z(x)>0$, other there would exist, by Lemma 1.2, $\lambda_1<\lambda$ such that the function $u_1=\lambda_1 v$ intersects the segment of z over $(\alpha_{n+1}, \alpha_{n+2})$ twice. That would mean that the function $w_1=z-\lambda_1 v$ has a double zero to the left of α_{n+1} while it has two zeros in $(\alpha_{n+1}, \alpha_{n+2})$. This, however, contradicts the assumption that the differential equation (b) is in class II. It is easy to verify that the function $w=z-\lambda v$ has $n+2$ zeros in $\langle\alpha_1, \xi\rangle$, including a double zero at ξ. Therefore, w is the principal solution of (b) determining the n-th conjugate point to $\alpha=\alpha_1$, while $\eta_n=\xi$ satisfies $\alpha_{n+1}<\xi<\alpha_{n+2}$, which completes the proof. □

THEOREM 2.30. If the differential equation (b) is in class II in (a, b), then it is oscillatory in (a, b) if and only if to each $\alpha \in (\bar{a}, b)$, $a < \bar{a} < b$ there exist infinitely many conjugate points $\eta_n(\alpha)$.

PROOF. Necessity. By definition, the differential equation (b) is oscillatory in (a, b) if there exists at least one solution of (b) which has infinitely many zeros in $(\bar{a}, b)$ with a limit point at b. That means that if z is an oscillatory solution of (b), there exists, for each α and for every positive integer n, $\bar{a} < \alpha < b$, a point $\beta > \alpha \in (\bar{a}, b)$ such that z has n zeros in (α, β). Let u be a non-trivial solution of the differential equation (b) such that $u(\alpha) = u(\beta) = 0$; then by Theorem 2.27, u has at least n zeros in $\langle \alpha, \beta \rangle$. Since n is arbitrary, it follows from Theorem 2.29 that there exist infinitely many points conjugate to α.

Sufficiency. Suppose that there exist infinitely many points conjugate to α. We have to show that there is a solution of the differential equation (b) which has infinitely many zeros in (α, b). We know that every solution z of (b) with $z(\alpha) = 0$ can be written in the form $z = c_1 u + c_2 v$, where u and v are independent solutions of (b) such that $u(\alpha) = u'(\alpha) = 0$, $u''(\alpha) \neq 0$, $v(\alpha) = v''(\alpha) = 0$, $v'(\alpha) \neq 0$ and c_1, c_2 are arbitrary constants. The solutions z can be normed by $c_1^2 + c_2^2 = 1$. By similar reasoning to that used when we defined conjugate points, we easily see that there is a subsequence $\{z_\nu(x)\}$ of solutions, which converges uniformly in every closed subinterval of (α, b) to a solution $z_0(x)$ of the differential equation (b).

According to Theorem 2.29, all the solutions z_ν, $\nu > n$, have a zero between η_{n-1}, η_n. Since every limit point of zeros of solutions z_ν from the sequence $\{z_\nu\}$ is a zero of z_0, the solution $z_0(x)$ must have a zero between any two neighbouring conjugate points. Thus the theorem is proved. □

In the above theorems we have shown that whenever the differential equation (a) is in class I, or the differential equation (b) is in class II, then they are oscillatory if and only if there exist infinitely many conjugate points in some interval $(\bar{a}, b)$, $a < \bar{a} < b$. This fact together with Theorem 2.20 and Lemma 2.11 establishes the following theorem.

THEOREM 2.31. The differential equation (a) in class I (or in class II) is oscillatory in (a, b) if and only if its adjoint equation (b) is oscillatory in (a, b). □

Theorem 2.31 enables us to establish a comparison theorem, analogous to Theorem 2.23, for the differential equation (b) compared with the differential equation

$$z''' + 2Az' + (A' - b_1)z = 0 \,. \tag{b_2}$$

THEOREM 2.32. Let the differential equations (b) and (b_2) be in class II in (a, b), let $b_1(x) \geqq b(x)$ for $x \in (a, b)$ and assume that $b_1(x) \not\equiv b(x)$ in any subinterval of (a, b). If the differential equation (b) is oscillatory in (a, b), then also the differential equation (b_2) is oscillatory in (a, b). □

8. *Criteria for Oscillatoricity of the Differential Equations (a) and (b) Implied by Properties of Conjugate Points*

The results presented in this section were derived by Hanan [61]. First of all, we shall generalize the comparison theorems from the preceding Section 7. We shall compare the differential equation (a) with the differential equation (a_1). Again, we shall assume that $b = \infty$, i.e. study the properties of solutions of these two equations in (a, ∞). We begin with proving the following lemma.

LEMMA 2.12. Let $A(x) \geqq 0$ and let $b(x)$ have the property (v) in (a, ∞). Let y be a solution of the differential equation (a) with $y(\alpha) = y'(\beta) = 0$, $a < \alpha < \beta < \infty$. Then y has another zero in the interval (β, ∞).

PROOF. It follows from the integral identity (1.6) for the solution y that

$$y'^2(x) = y'^2(\alpha) + 2y(x)y''(x) + 2A(x)y^2(x) + \\ + 2\int_\alpha^x b(t)y^2(t)\,dt \,. \tag{2.37}$$

Assuming that $y(x)>0$ for $x>\alpha$, we deduce from (2.37) that $y(x)y''(x)<0$ at every x with $y'(x)=0$. However, $y'(x)=0$ can only happen once in (α, ∞). Here, one of the following three cases may arise.

a) $y''(x)<0$ for $x>\beta$,

b) $y''(x)>0$ for $x>\beta$,

c) $y''(x)$ has infinitely many zeros for $x>\beta$.

We are going to show that, in all three cases, y has at least one zero for $x>\beta$. This is evident in case a). In fact, if two consecutive derivatives of y are negative, then necessarily y has a zero at sufficiently great x. In the case b), observe that the right-hand side of (2.37) is positive since $y'^2(x)>0$ for $x>b$. Hence $y'^2(x)$ converges to a positive constant $k^2\neq 0$, because the last term in (2.37) is an increasing function of x. Therefore, $y'(x)\to -k\neq 0$ as $x\to\infty$. This again implies that y must intersect the y axis. In the case c), let x approach infinity, through points at which $y'(x)=0$, and a similar conclusion as in case b) follows. Thus, the theorem is proved. □

THEOREM 2.33. Suppose that the coefficients of the differential equations (a) and (a_1) satisfy

$$A_1(x)\geqq A(x)\geqq 0,\quad A_1'(x)+b_1(x)\geqq A'(x)+b(x) \tag{2.38}$$

for $x\in(y, \infty)$, and let $b(x)$ and $b_1(x)$ have the property (v) in (a, ∞).

If $\eta_n(\alpha)$ and $\eta_n'(\alpha)$, $a<\alpha<\infty$, are the n-th conjugate points of (a) and (a_1), then

$$\eta_n'(\alpha)\leqq \eta_{(k+1)n-1}(\alpha) \tag{2.39}$$

for some positive integer k.

PROOF. Let $u(x)=y(x, \alpha)$ and $v(x)=y_1(x, \alpha)$ be the first principal solutions of the differential equations (a) and (a_1) and let $\beta=\eta_1(\alpha)$ be the first zero of $u(x)$. By Theorem 2.20, a principal solution $w(x)$ of the differential equation adjoint to (a), i.e. of the differential equation (b), which has a zero at α, has a double zero at β and has no zero in (α, β). Without loss of generality we may assume that $w(x)>0$ in (α, β). If we multiply the differential equation (a) by $v(x)$ and the differential equation (a_1) with $z=v(x)$ by $w(x)$, then subtract and integrate from α to β, we obtain

$$[wv''-w'v'+w''v]_\alpha^\beta+\int_\alpha^\beta[2A_1wv'+2Aw'v+2A'wv]\,dt+$$

$$+\int_\alpha^\beta(A_1'+b_1-A'-b)wv\,dt=0\,,$$

or

$$w''(\beta)v(\beta)+\int_\alpha^\beta(2A_1-2A)wv'\,dt+$$

$$+\int_\alpha^\beta(A_1'+b_1-A'-b)wv\,dt=0\,. \tag{2.40}$$

By hypothesis $w''(\beta)>0$, while $v(x)$ and $v'(x)$ are positive in a right neighbourhood of α. There are three cases to be considered

a) both $v(x)$ and $v'(x)$ are positive for all x,
b) $v(x)>0$, but $v'(x)$ has a zero in $\langle\alpha,\beta\rangle$,
c) $v(x)$ has a zero in $\langle a,\beta\rangle$.

It is obvious that the case a) cannot arise, because the sum on the left-hand side of (2.40) is positive. Under the conditions of case b), Lemma 2.12 guarantees that $v(x)$ has a zero at some $\beta=\eta_1'(\alpha)$. If that point lies in the interval $\langle\eta_{k-1}(\alpha),\ \eta_k(\alpha)\rangle$, i.e. if

$$\eta_{k-1}(\alpha)\leqq\beta_1\leqq\eta_k(\alpha)\,, \tag{2.41}$$

then the rest of the proof of Theorem 2.33 is analogous to that of Theorem 2.23. Of course, the integer k in (2.41) is positive as in (2.39).

It is evident that in case c) the assertion of Theorem 2.33 is the same as that of Theorem 2.23, that is, (2.39) can be replaced by the stronger inequality (2.32). Thus the theorem is proved. □

REMARK 2.18. Although the assertion of Theorem 2.33 is not as strong as that of Theorem 2.23, yet Theorem 2.33 guarantees that whenever there exist infinitely many conjugate points of the differential equation (a), the same is true of the differential equation (a_1).

Remarks 2.17 and 2.18 imply the following theorem.

THEOREM 2.34. If the differential equation (a) is oscillatory and if the inequalities (2.38) hold true and the functions $b(x)$ and $b_1(x)$ have the property (v) in (a, ∞), then also the equation (a_1) is oscillatory. □

For the differential equation (b) and the differential equation

$$z''' + 2A_1 z' + (A_1' - b_1)z = 0 \tag{b$_1$}$$

adjoint to (a_1), Theorem 2.34 can be stated in the following formulation.

THEOREM 2.35. Let the coefficients of differential equations (b) and (b_1) satisfy

$$A_1(x) \geqq A(x) \geqq 0, \quad A_1'(x) - b_1(x) \geqq A'(x) - b(x) \tag{2.42}$$

and let the functions $b_1(x)$ and $b(x)$ have the property (v) in (a, ∞). If the differential equation (b) is oscillatory in (a, ∞), then so is (b_1). □

Comparison criteria make it possible to derive criteria for oscillatoricity of solutions of the differential equation (a) if oscillatoricity properties of a particular given differential equation are known. As an example, consider Euler's differential equation

$$y''' + \frac{m}{x^2} y' + \frac{n}{x^3} y = 0 , \tag{2.43}$$

where m and n are constants, $x \in (0, \infty)$.

It is not difficult to prove that if $m > 1$, then the equation (2.43) is oscillatory for all n. If $m < 1$, it is necessary to determine the sign of the Laguerre invariant $b = b(x)$ in order to apply Theorem 2.34. For the differential equation (2.43) we have $b(x) = x^{-3}(m+n)$ and we can prove that if $m + n > 0$, then the differential equation (2.43) is oscillatory in $(0, \infty)$ if and only if $m + n - 2[(1-m)/3]^{3/2} > 0$. If $m + n < 0$, then the differential equation (2.43) is oscillatory in $(0, \infty)$ if and only if $m + n + 2[(1-m)/3]^{3/2} < 0$. These remarks and Theorem 2.34 are sufficient to establish the following results.

THEOREM 2.36. If $b(x)$ has the property (v) in (a, ∞) and if there exists a number α such that

$$\liminf_{x \to \infty} 2x^2 A(x) > v > 1, \quad \liminf_{x \to \infty} x^3[A'(x) + b(x)] > -\alpha ,$$

then the differential equation (a) is oscillatory in (a, ∞). □

THEOREM 2.37. If the function $b(x)$ has the property (v) in (a, ∞) and if there exists a positive $\alpha < 1$ with

$$\liminf_{x\to\infty} 2x^2 A(x) > \alpha, \quad \liminf_{x\to\infty} x^3[A'(x)+b(x)] > \\ > 2\left(\frac{1-\alpha}{3}\right)^{3/2} - \alpha,$$

then the differential equation (a) is oscillatory in (a, ∞). If

$$\limsup_{x\to\infty} 2x^2 A(x) < \alpha \leqq 1, \quad \limsup_{x\to\infty} x^3[A'(x)+b(x)] < \\ < 2\left(\frac{1-\alpha}{3}\right)^{3/2} - \alpha,$$

then (a) is non-oscillatory in (a, ∞). □

If $A(x) \equiv 0$ for $x \in (a, \infty)$, the differential equation (a) becomes a binomial equation of the form

$$y''' + b(x)y = 0 . \tag{2.44}$$

Comparison of the last equation with Euler's differential equation (2.43), in which we put $m = 0$, gives the following result.

THEOREM 2.38. If the function $b(x)$ has the property (v) in (a, ∞) and

$$\liminf_{x\to\infty} x^3 b(x) > \frac{2}{3\sqrt{3}},$$

then the differential equation (2.44) is oscillatory in (a, ∞).

If

$$\limsup_{x\to\infty} x^3 b(x) < \frac{2}{3\sqrt{3}},$$

then the differential equation (2.44) is non-oscillatory in (a, ∞). □

REMARK 2.19. The results contained in Theorem 2.38 are equivalent with those in Corollary 2.7.

Now we make use of Theorem 2.15 and prove the following theorem.

THEOREM 2.39. Let the second order differential equation

$$y''+2A(x)y=0 \tag{2.45}$$

be disconjugate in (a, ∞). Moreover, let

$$2A' \leqq A' + b \leqq 0 \,. \tag{2.46}$$

Then the differential equation (a) is disconjugate in (a, ∞).

PROOF. From (2.46), we have $A' - b \leqq 0$. Then by Remark 2.15, the differential equation (b) is in class II. Lemma 2.11 implies that its adjoint equation (a) belongs to class I. Therefore, no solution of the differential equation (a) with a double zero at α, $a<\alpha<\infty$, has another zero to the left of α. On the other hand, from (2.46) and Remark 2.15 it follows that the differential equation (a) is in class II. Hence no solution with a double zero at α can have another zero to the right of α. The assertion of Theorem 2.39 now follows from properties of bands. □

The following two theorems give criteria for oscillatoricity of the differential equation (a), which depend on integrability of the functions $x[b(x)-A'(x)]$ and $x^2[b(x)-A'(x)]$.

THEOREM 2.40. Let the differential equation (2.45) be disconjugate in (a, ∞) and let the functions $A(x)$, $A'(x)+b(x)$, $b(x)-A'(x)$ be positive in (a, ∞). If

$$\int_\alpha^\infty t[b(t)-A'(t)]\,\mathrm{d}t=\infty\,, \quad a<\alpha<\infty\,,$$

then the differential equation (a) is oscillatory in (a, ∞).

PROOF. Suppose the contrary, i.e. that the differential equation (a) is non-oscillatory in (a, ∞). Since $A'(x)+b(x)>0$, the differential equation (a) is in class I, thus it follows from the properties of bands that, starting from some $x_0>a$, every solution of (a) has at most two zeros in (x_0, ∞). Hence the first principal solution $u(x)=y(x, x_0)$ is positive for $x>x_0$. We shall show that $u'(x)>0$ for $x>x_0$.

Suppose this is not true. Then, since $u'(x)>0$ in a sufficiently small right neighbourhood of x_0, there exists a point $x_1 > x_0$ with $u'(x_1) = 0$.

By Remark 2.15, $u(x)$ has no other zeros for $x > x_1$, so $u'(x) < 0$ for $x > x_1$ and $u'(x_1) \leqq 0$. If two consecutive derivatives of $u(x)$ were negative for $x > x_1$, we would have $u(x) < 0$ from some x onwards. This is impossible, hence $u''(x)$ must be positive beginning with some x. There exists therefore a point $c \geqq x_1$ with $u''(c) = 0$. Multiply the differential equation (a) by y', integrate from x_1 to c and put $y = u$. We get

$$\int_{x_1}^{c} 2Au'^2 \, dt + \int_{x_1}^{c} (A' + b)uu' \, dt = \int_{x_1}^{c} u''^2 \, dt . \qquad (2.47)$$

Nehari has proved in his paper [104] that if $v(x)$ is any function having a piecewise continuous first derivative in $\langle \alpha, \beta \rangle$ with $v'(\alpha) = 0$ and if the differential equation (2.45) is disconjugate in $\langle \alpha, \beta \rangle$, then

$$\int_{\alpha}^{\beta} v'^2 \, dt > \int_{\alpha}^{\beta} 2Av^2 \, dt , \quad a < \alpha < \beta .$$

Setting $v = u'$ in the last inequality, a contradiction will arise from (2.47), proving that necessarily $u'(x) > 0$ for $x > x_0$. Using this fact, we deduce from the differential equation (a) that $u'''(x) < 0$ for $x > x_0$, and so $u''(x)$ is a decreasing function for $x > x_0$. On the other hand, $u''(x)$ cannot be negative, because otherwise it would follow from $u'''(x) < 0$, $u''(x) < 0$ that u intersects the x axis. Therefore, $u''(x) > 0$ and hence $u'(x)$ is an increasing function of x for $x > x_0$. Thus $u'(x) > 0$ for $x \in (x_0, x_1\rangle$ and $u'(x) > \bar{A} > 0$ for $x \geqq x_1$. Due to $u(x_1) > 0$, we have

$$u(x) = u(x_1) + (x - x_1)u'(\xi) > (x - x_1)\bar{A}, \qquad x_1 < \xi < x .$$

From the integral identity (1.7) for the solution u we obtain

$$u''(x_1) + 2A(x_1)u(x_1) = u''(x) + 2A(x)u(x) +$$
$$+ \int_{x_1}^{x} (b - A')u(t) \, dt \geqq \int_{x_1}^{x} (b - A')u(t) \, dt \geqq$$
$$\geqq \bar{A} \int_{x_1}^{x} [b(t) - A'(t)](t - x_1) \, dt.$$

This implies

$$\int_{x_1}^{\infty} (t-x_1)\,[b(t)-A'(t)]\,\mathrm{d}t<\infty\,,$$

which contradicts the assumption. The theorem is thus proved. □

THEOREM 2.41. Let the differential equation (2.45) be disconjugate in (a, ∞) and let the functions $A(x)$, $A'(x)+b(x)$, $b(x)-A'(x)$ be positive in (a, ∞). If

$$\int_{a}^{\infty} t^2[b(t)-A'(t)]\,\mathrm{d}t<\infty\,, \tag{2.48}$$

then the differential equation (a) is non-oscillatory in (a, ∞).

PROOF. Integrating the differential equation (a) three times from α to x, $a<\alpha<\infty$, $a<x<\infty$, yields

$$y(x)=y(\alpha)+(x-\alpha)y'(\alpha)+\frac{1}{2}(x-\alpha)^2\Big[y''(\alpha)+$$

$$+2A(\alpha)y(\alpha)\Big]-\int_{\alpha}^{x}(x-t)2A(t)y(t)\,\mathrm{d}t-$$

$$-\frac{1}{2}\int_{\alpha}^{x}(x-t)^2[b(t)-A'(t)]y(t)\,\mathrm{d}t\,.$$

From the above equality we obtain the first principal solution $u(x)=y(x, \alpha)$ of the differential equation (a) in the form

$$u(x)=\frac{1}{2}(x-\alpha)^2-2\int_{\alpha}^{x}(x-t)A(t)u(t)\,\mathrm{d}t-$$

$$-\frac{1}{2}\int_{\alpha}^{x}(x-t)^2[b(t)-A'(t)]u(t)\,\mathrm{d}t\,. \tag{2.49}$$

Assume that the differential equation (a) is oscillatory in (a, ∞). The first principal solution is oscillatory in (a, ∞) in view of the fact that the equation (a) is in class I. Let $x=\beta$ be the first zero of the solution $u(x)$, $\beta>\alpha$. Since $u(x)>0$ in (α, β), we get from (2.49) that

$$u(x)\leqq\frac{1}{2}(x-\alpha)^2, \qquad a<x<b\,. \tag{2.50}$$

After substituting $x=\beta$ in (2.49) we have

$$\frac{1}{2}(\beta-\alpha)^2=2\int_\alpha^\beta(\beta-t)A(t)u(t)\,dt+$$
$$+\frac{1}{2}\int_\alpha^\beta(\beta-t)^2[b(t)-A'(t)]u(t)\,dt\leqq$$
$$\leqq(\beta-\alpha)2\int_\alpha^\beta A(t)u(t)\,dt+$$
$$+\frac{1}{2}(\beta-\alpha)^2\int_\alpha^\beta[b(t)-A'(t)]u(t)\,dt\,,$$

or

$$1\leqq\int_\alpha^\beta[b(t)-A'(t)]u(t)\,dt+\frac{4}{\beta-\alpha}\int_\alpha^\beta A(t)u(t)\,dt\,.$$

Making use of (2.50), in the last inequality we get

$$1\leqq\frac{1}{2}\int_\alpha^\beta(t-\alpha)^2[b(t)-A'(t)]\,dt+$$
$$+\frac{2}{\beta-\alpha}\int_\alpha^\beta(t-\alpha)^2A(t)\,dt\,. \tag{2.51}$$

Let us try to estimate the second integral in (2.51), using a result from a paper by Nehari [104] which states that if the differential equation (2.45) is non-oscillatory and $A(x)\geqq 0$, then

$$\frac{2}{\beta-\alpha}\int_\alpha^\beta(t-\alpha)^2A(t)\,dt<1-2(\beta-\alpha)\int_\beta^\infty A(t)\,dt. \tag{2.52}$$

Two cases must be considered:

a) $\limsup_{x\to\infty} x\int_x^\infty 2A(t)\,dt=\gamma\neq 0,\quad \gamma<1,$

b) $\limsup_{x\to\infty} x\int_x^\infty 2A(t)\,dt=0.$

We shall show that in either case the inequality (2.51) gives rise to a contradiction.

Investigate the case a) first. In (2.48), choose α sufficiently large to make

$$\int_{\alpha}^{\infty} t^2[b(t)-A'(t)]\,dt \leqq \gamma\,. \tag{2.53}$$

Substituting from (2.52) in (2.51) we get

$$(\beta-\alpha)2\int_{\beta}^{\infty} A(t)\,dt < \frac{1}{2}\int_{\alpha}^{\beta} (t-\alpha)^2[b(t)-A'(t)]\,dt <$$

$$< \frac{1}{2}\int_{\alpha}^{\beta} t^2[b(t)-A'(t)]\,dt\,.$$

If we let β tend to infinity through a sequence of points for which the function $2\beta\int_{\beta}^{\infty} A(t)\,dt$ attains its upper limit, the last equality yields

$$2\gamma < \int_{\alpha}^{\beta} t^2[b(t)-A'(t)]\,dt\,,$$

contradicting (2.53). Thus, in the case a), the differential equation (a) is non-oscillatory.

In the case b), put

$$r(t) = (t-\alpha)\int_{t}^{\infty} p(x)\,dx$$

and integrate by parts the second summand on the right-hand side of (2.51). We obtain

$$\frac{1}{\beta-\alpha}\int_{\alpha}^{\beta}(t-\alpha)^2 2A(t)\,dt = (\beta-\alpha)\int_{\alpha}^{\beta} 2A(x)\,dx -$$

$$-\frac{2}{\beta-\alpha}\int_{\alpha}^{\beta}(t-\alpha)\left(\int_{t}^{\infty} 2A(x)\,dx\right)dt =$$

$$= (\beta-\alpha)\int_{\beta}^{\infty} 2A(x)\,dx - \frac{2}{\beta-\alpha}\int_{\alpha}^{\beta} r(t)\,dt\,.$$

Letting β tend to infinity along a sequence of points for which $\beta\int_{\beta}^{\infty} 2A(x)\,dx$ attains its upper limit, we deduce from the last equality that

$$\lim_{\beta\to\infty}\left[\frac{1}{\beta-\alpha}\int_\alpha^\beta (t-\alpha)^2 2A(t)\,dt\right]=0\,.$$

We use this result and let β to infinity in (2.51); this gives

$$2\leqq\int_\alpha^\infty (t-\alpha)^2[b(t)-A'(t)]\,dt<\int_\alpha^\beta t^2[b(t)-A'(t)]\,dt\,,$$

which again contradicts (2.53). The theorem is thus proved. □

For the differential equation (b), the preceding two theorems may be restated as follows.

THEOREM 2.42. Let $A(x)>0$, $A'(x)+b(x)\leqq 0$, $A'(x)-b(x)\geqq 0$ in (a, ∞). If

$$\int_\alpha^\infty t[A'(t)+b(t)]\,dt=-\infty\,,\quad a<\alpha<\infty\,,$$

then the differential equation (b) is oscillatory in (a, ∞).

If

$$-\int_\alpha^\infty t^2[A'(t)+b(t)]\,dt<\infty\,,\quad a<\alpha<\infty\,,$$

then the differential equation (b) is non-oscillatory in (a, ∞). □

A question arises as to what happens with non-oscillatory solutions of the differential equation (a) when the second order equation (2.45) is disconjugate. We shall prove the following theorems (Hanan [61], Lazer [86]).

THEOREM 2.43. Let the equation (2.45) be disconjugate in (a, ∞) and let $A'(x)+b(x)>0$ for $x\in(a, \infty)$. If y is a solution of the differential equation (a) such that y' is an oscillatory function in (a, ∞), then the solution y oscillates in (a, ∞).

PROOF. By virtue of (2.47), if y has no zero between two zeros of y' then y and y' have the same sign in the interval under consideration. Assume that $y(x)>0$ in (a, ∞), then necessarily $y'(x)\geqq 0$ in (a, ∞). We show that this is not possible. Let $\alpha\in(a, \infty)$ be a point at which $y'(\alpha)=y''(\alpha)=0$. Then $y'(x)$ would have to be negative in some neighbourhood of α, because it follows from the differential equation (a) that $y'''(\alpha)<0$. Thus, the theorem is proved. □

THEOREM 2.44. Let the equation (2.45) be disconjugate in (a, ∞) and let $A'(x)+b(x)>0$ for $x\in(a, \infty)$. Let $u(x)=y(x, \alpha)$ and $v(x)=y(x, \beta)$ be two principal solutions of the differential equation (a) with $\beta>\alpha\in(a, \infty)$. If sgn $u(\beta)=$ sgn $u'(\beta)$, then the zeros of u and v separate each other in (β, ∞).

PROOF. Suppose that $u(\beta)>0$, $u'(\beta)>0$. We know that $\sigma=uv'-u'v$ is a solution of the differential equation (b), while

$$\sigma(\beta)=0\,,\quad \sigma'(\beta)=u(\beta)v''(\beta)>0\,, \qquad (2.54)$$
$$\sigma''(\beta)=u'(\beta)v''(\beta)>0\,.$$

We shall show that $\sigma(x)>0$ for $x>\beta$. Suppose this is not true. Then there exists a point $c>\beta$ with $\sigma(c)=0$. Let w be a solution of the differential equation (a) with $w(c)=w'(c)=0$, $w''(c)>0$. Multiply the differential equation (a), in which $y=w$, by the function σ, and the differential equation (b), in which $z=\sigma$, by the function w. Then, after summing and integrating from β to c, we get

$$\sigma'(\beta)w'(\beta)-\sigma''(\beta)w(\beta)=0\,. \qquad (2.55)$$

Since w has a double zero at c and $w''(c)>0$, Remark 2.15 implies that $w(x)>0$ and $w'(x)<0$ for $x<c$. From the property (2.54) of σ at β it follows that the left-hand side of (2.55) is negative, which is a contradiction. Thus, $\sigma(x)>0$ for $x>\beta$. Therefore, the zeros of the first principal solutions u and v separate each other in the interval (β, ∞), and the theorem is proved. □

THEOREM 2.45. Let the equation (2.45) be disconjugate in (a, ∞) and assume that $A'(x)+b(x)>0$ in (a, ∞). If the differential equation (a) is oscillatory in (a, ∞) and $u(x)$ is any non-oscillatory solution of (a), then $u(x)$ is a decreasing function from some $x_0>a$ on.

PROOF. It follows from the properties of bands that u cannot have a zero in (a, ∞), otherwise it would be an oscillatory solution in (a, ∞).

Suppose that $u(x)>0$ in (a, ∞) and that u is not a decreasing function of $x\in(x_0, \infty)$, $x_0>a$, i.e. that $u'(x)>0$ in some interval (α, ∞). According to Theorem 2.44, u cannot have two zeros in (a, ∞). Let $v(x)=y(x, \alpha)$ be the first principal solution of the differential equation (a). The solution is therefore oscillatory in (a, ∞). Let $\beta_1>\alpha$ be the first zero of v to the right of α; then $v(x)>0$ in (α, β_1). By Lemma 1.2, there exists a solution $w(x)=v(x)-\lambda u(x)$ having a double zero at some $\xi\in(\alpha, \beta_1)$. Since u and v are positive in (α, β_1), we have $\lambda>0$, and $u'(\xi)>0$ implies also that $v'(\xi)>0$. According to Theorem 2.44, the zeros of the solutions v and w separate each other in (ξ, ∞). This, however, is impossible since $\lambda u(x)$ intersects every positive segment of the function v twice. Therefore, $u(x)$ cannot be positive in (α, ∞). This completes the proof. □

Using Theorem 2.45, we shall prove the following test for non-oscillatoricity of the differential equation (a).

THEOREM 2.46. Let $A(x)\geqq 0$, $A'(x)+b(x)>0$, let $b(x)$ have the property (v) in (a, ∞) and assume $\lim_{x\to\infty} A(x)=0$ and $\int_a^\infty [A'(t)+b(t)]\,dt<\infty$. If the second order differential equation

$$y''+\left[2A(x)+\frac{3}{2}\int_x^\infty [A'(t)+b(t)]\,dt\right]y=0 \tag{2.56}$$

is disconjugate in (a, ∞), then the differential equation (a) is non-oscillatory in (a, ∞).

PROOF. We are going to prove that if the hypotheses, except for the disconjugateness of (2.56), are satisfied and if the differential equation (a) has an oscillatory solution, then the differential equation (2.56) is oscillatory. If the equation (2.45) were oscillatory, then evidently (by the Sturm comparison theorem for second order equations – Sansone [124]) the differential equation (2.56) would be oscillatory as well. Suppose, therefore, that the differential equation (2.56) is disconjugate in (a, ∞). Theorem 2.45 implies that the absolute value of any solution $u(x)$ without zeros is a decreasing function of $x\in\langle x_0, \infty)$. Let $u(x)>0$, then $\lim_{x\to\infty} u(x)=c\geqq 0$. Thus, for every $\varepsilon>0$, we evidently have

$$0 \leqq u''(x) < \varepsilon \tag{2.57}$$

for sufficiently great $x \in (a, \infty)$. Since $\lim_{x \to \infty} A(x) = 0$, for sufficiently great x we have

$$0 \leqq 2A(x) < \varepsilon\,. \tag{2.58}$$

On the other hand, for sufficiently great x one has

$$0 < u(x) < c + 1\,. \tag{2.59}$$

From (2.57), (2.58) and (2.59) there follows, for sufficiently great x, the inequality

$$\begin{aligned} -2F(u(x)) &= u'^2(x) - 2u(x)u''(x) - 2A(x)u^2(x) \geqq \\ &\geqq -2\varepsilon(c+1) - \varepsilon(c+1)^2\,. \end{aligned} \tag{2.60}$$

As $-2F(u(x))$ is always negative and increasing (cf. Theorem 2.17), it follows from (2.60) that $\lim_{x \to \infty} F(u(x)) = 0$, or

$$u'^2(x) - 2u(x)u''(x) - 2A(x)u^2(x) = -2 \int_x^\infty b(t)u^2(t)\,dt\,.$$

Since $u(x) \neq 0$ for $a < x < \infty$, it follows that

$$\frac{2u''(x)}{u(x)} - \frac{u'(x)}{u^2(x)} = -2A(x) + \frac{4}{u^2(x)} \int_x^\infty b(t)u^2(t)\,dt\,. \tag{2.61}$$

Since $u(x)$ is a decreasing function and $\lim_{x \to \infty} A(x) = 0$, we have

$$\begin{aligned} \frac{1}{u^2(x)} \int_x^\infty 2b(t)u^2(t)\,dt &\leqq \int_x^\infty 2b(t)\,dt = \\ &= 2\int_x^\infty [A''(t) + b(t)]\,dt + 2A(x)\,. \end{aligned}$$

Thus, (2.61) implies that

$$\frac{2u''(x)}{u(x)}-\frac{u'^2(x)}{u^2(x)}\leqq 2\int_x^\infty [A'(t)+b(t)]\,dt \qquad (2.62)$$

for all $x\in(a, \infty)$.

Let $y(x)=u(x)\int_\alpha^\infty z(t)\,dt$ with z chosen so that y is a solution of the differential equation (a). Substituting in (a), we get the following equation for z

$$z''+\frac{3u'}{u}z'+\left[2A+\frac{3u''}{u}\right]z=0\,,$$

which by substitution $z=wu^{-3/2}$ becomes

$$w''+\left[2A+\frac{3}{4}\left(\frac{2u''}{u^2}-\frac{u'^2}{u^2}\right)\right]w=0\,. \qquad (2.63)$$

Every non-zero solution w of (2.63) must be oscillatory, because

$$y(x)=u(x)\int_\alpha^x u(t)^{-3/2}w(t)\,dt$$

is a non-trivial oscillatory solution of the differential equation (a) and $y(\alpha)=0$. Since it follows from (2.62) that

$$2A(x)+\frac{3}{4}\left[\frac{2u''(x)}{u(x)}-\frac{u'^2(x)}{u^2(x)}\right]\leqq$$
$$\leqq 2A(x)+\frac{3}{2}\int_x^\infty [A'(t)+b(t)]\,dt\,,$$

we conclude from the Sturm comparison theorem that whenever the differential equation

$$y''+\left[2A+\frac{3}{4}\left(\frac{2u''}{u}-\frac{u'^2}{u^2}\right)\right]y=0$$

is oscillatory, then the equation

$$y''+\left[2A(x)+\frac{3}{2}\int_x^\infty [A'(t)+b(t)]\,dt\right]y=0$$

is also oscillatory in (a, ∞). The theorem is thus proved. □

REMARK 2.20. In Gera's paper [28], some further relations are studied between oscillation properties of the differential equation (a) and the second order differential equation of the form

$$u'' + \left[\frac{1}{2} A(x) - \frac{3}{2} \inf_{\xi \in I} \int_{\xi}^{x} b(t)\, dt\right] u = 0 ,$$

or of the form

$$u'' + \left[\frac{1}{2} A(x) + \frac{3}{2} \sup_{x \in I} \int_{\xi}^{x} b(t)\, dt\right] u = 0 ,$$

where $I = \langle x_0, x)$, $a < x_0 < x < \infty$. However, we shall not analyse these results here.

9. *Further Criteria for Oscillatoricity of the Differential Equation (b)*

By Lemma 1.1, given any solution w of the differential equation (b), there exist two independent solutions y_1 and y_2 of the differential equation (a) such that $w(x) = y_1(x)y_2'(x) - y_1'(x)y_2(x)$ for $x \in (a, b)$. The converse of Lemma 1.1 is evidently also true, i.e. for any solution y of the differential equation (a) there exist two independent solutions z_1, z_2 of the differential equation (b) satisfying $y(x) = z_1(x)z_2'(x) - z_1'(x)z_2(x)$ for $x \in (a, b)$.

Throughout this section, we shall again assume that $(a, b) \equiv (a, \infty)$, that is, $b = \infty$. Theorem 1.13 implies that if the function $b(x)$ has the property (v) in (a, ∞), then there is at least one solution of the differential equation (a) with no zeros in (a, ∞).

From the above remarks it follows that whenever the function $b(x)$ has the property (v) in (a, ∞), there exists a two-parameter system (band) of solutions $z = c_1 z_1 + c_2 z_2$ of the differential equation (b) satisfying, on the whole interval (a, ∞), the differential equation (c) with $w = Y$, i.e. the following second order differential equation

$$Yz'' - Y'z' + (Y'' + 2AY)z = 0 .$$

This band has the property that the zeros of the solutions (if they exist) separate each other. Y is a solution of (a) without zeros.

The aim of this section is to prove that, under certain assumptions about the coefficients of the differential equation (b), the zeros of any two independent solutions separate each other in (a, ∞) (Lazer [86]).

We shall also give several other results concerning oscillatory properties of solutions of the differential equation (b) (Jones [74]).

LEMMA 2.13. Assume that $A(x) \leqq 0$, $A'(x) - b(x) \leqq 0$ for $x \in (a, \infty)$ and let z be any solution of the differential equation (b) with

$$z(x_0) \geqq 0, \quad z'(x_0) \geqq 0, \quad z''(x_0) > 0, \quad a < x_0 < \infty .$$

Then

$$z(x) > 0, \quad z'(x) > 0, \quad z''(x) > 0, \quad z'''(x) \geqq 0 \quad \text{for } x > x_0$$

and, moreover, $\lim_{x \to \infty} z(x) = \lim_{x \to \infty} z'(x) = \infty$.

PROOF. First of all, we show that $z''(x) > 0$ for $x \geqq x_0$. Consider the function

$$V(x) = z(x) z'(x) z''(x).$$

If $z''(x)$ has any zeros for $x > x_0$, there exists the least $x_1 > x_0$ with $z''(x_1) = 0$. From $z(x_0) \geqq 0$, $z'(x_0) \geqq 0$, $z''(x_0) > 0$ it follows that $z(x) > 0$, $z'(x) > 0$ for $x \in (x_0, x_1)$, $V(x_0) \geqq 0$, $V(x_1) = 0$. Since $A(x) \leqq 0$ and $A'(x) - b(x) \leqq 0$ for $x \in (a, \infty)$, we have

$$\frac{\mathrm{d}V(x)}{\mathrm{d}x} = (z''(x))^2 z(x) + z''(x)(z'(x))^2 - 2A(x)\,(z'(x))^2 - \\ -[A'(x) - b(x)] z^2(x) z'(x) > 0$$

for $x \in (x_0, x_1)$.

Integration of the last equality from x_0 to x_1 yields a contradiction, i.e.

$$0 = V(x_0) + \int_{x_0}^{x} V'(t)\,\mathrm{d}t > 0 .$$

Therefore $z''(x) > 0$ for $x \geqq x_0$, and since $z(x_0) \geqq 0$, $z'(x_0) \geqq 0$ we have $z(x) > 0$, $z'(x) > 0$ and $z''(x) > 0$ for $x > x_0$. Finally, $z'''(x) = -2A(x) z'(x) - [A'(x) - b(x)] z(x) \geqq 0$ for $x > x_0$. These inequalities clearly imply that $\lim_{x\to\infty} z(x)=\lim_{x\to\infty} z'(x)=\infty$. Thus, Lemma 2.13 is proved. □

LEMMA 2.14. If $A(x)\leqq 0$, $A'(x)-b(x)\leqq 0$ and $z(x)$ is any non-trivial, non-oscillatory solution of the differential equation (b), then there exists a number $c\in(a, \infty)$ such that either $z(x)z'(x)>0$ for $x>c$ or $z(x)z'(x)\leqslant 0$ for $x\geqq c$.

PROOF. By Lemma 2.13, if z is any non-trivial, non-oscillatory solution of the differential equation (b), then $z(x)$ can have at most one double zero. Without loss of generality we may assume that $z(x)>0$ for $x\geqq\beta$. To prove Lemma 2.14, it is sufficient to show that the sign of $z'(x)$ can change from negative to positive at most once in the interval $\langle\beta, \infty)$. Let c be a point at which $z(c)>0$, $z'(c)>0$ and $z''(c)>0$. According to Lemma 2.13, we have $z(x)>0$ and $z'(x)>0$ for $x>c$; we have thus proved the lemma. □

THEOREM 2.47. If $A(x)\leqq 0$, $A''(x)-b(x)\leqq 0$ and the differential equation (b) has an oscillatory solution, then for every non-trivial, non-oscillatory solution z there exists some $c\in(a, \infty)$ with

$$z(x)z'(x)z''(x)\neq 0, \quad \operatorname{sgn} z(x)=\operatorname{sgn} z'(x)=\operatorname{sgn} z''(x)$$

for $x\geqq c$ and, moreover,

$$\lim_{x\to\infty} |z(x)|=\lim_{x\to\infty} |z'(x)|=\infty .$$

PROOF. If z is any non-trivial, non-oscillatory solution of the differential equation (b), then by Lemma 2.14 there exists $d\in(a, \infty)$ such that either $z(x)z'(x)>0$ or $z(x)z'(x)\leqq 0$ for $x\geqq d$. Therefore, either $\lim_{x\to\infty} z(x)$ exists and is finite or $|z(x)|\to\infty$. Let $v(x)$ be an oscillatory solution of the differential equation (b). Construct the wronskian $W(v(x), z(x))=vz'-v'z$. Certainly, $W(v, z)$ has a zero at some $x\in(a, \infty)$, otherwise the zeros of v and z would separate each other, which would contradict the assumption that z is non-oscillatory. If β is a zero of the function $W(v, z)$, there exist non-zero constants c_1, c_2 such that

$$c_1 v(\beta) + c_2 z(\beta) = 0 \,,$$
$$c_1 v'(\beta) + c_2 z'(\beta) = 0$$

while

$$c_1 v''(\beta) + c_2 z''(\beta) > 0 \,.$$

Now consider the solution $u = c_1 v + c_2 z$.

Since $u(\beta) = u'(\beta) = 0$, $u''(\beta) > 0$, it follows from Lemma 2.12 that

$$\lim_{x\to\infty} u(x) = \lim_{x\to\infty} u'(x) = \infty \,. \tag{2.64}$$

We know that $\lim\limits_{x\to\infty} z(x)$ exists and either is finite or $z(x) \to \pm\infty$. If $\lim\limits_{x\to\infty} z(x)$ were finite, we would have

$$\lim_{x\to\infty} c_1 v(x) = \lim_{x\to\infty} [u(x) - c_2 z(x)] = \infty$$

and v could not be oscillatory. Therefore, $\lim\limits_{x\to\infty} z(x) = \pm\infty$, and it follows from Lemma 2.14 that there exists $c \in (a, \infty)$ such that $z(x) z'(x) > 0$ for $x \geqq c$. With no loss of generality we may assume that $z(x) > 0$ and $z'(x) > 0$ for $x \geqq c$, thus

$$z'''(x) = -2A(x) z'(x) - [A'(x) - b(x)] z(x) \geqq 0$$

for $x \geqq c$. The last inequality implies that for some $d_1 \geqq c$ we have either $z''(x) > 0$ or $z''(x) \leqq 0$ for all $x \geqq d_1$. In the latter case, $\lim\limits_{x\to\infty} z'(x)$ would be finite, since $z'(x) > 0$ for $x \geqq d \geqq c$. Then, however, (2.64) would imply that

$$\lim_{x\to\infty} c_1 v'(x) = \lim_{x\to\infty} [u'(x) - c_2 z'(x)] = \infty \,.$$

It follows that v cannot oscillate. Thus,

$$\operatorname{sgn} z(x) = \operatorname{sgn} z'(x) = \operatorname{sgn} z''(x)$$

for $x \geqq d_1$ and, at the same time,

$$\lim_{x\to\infty} | z(x) | = \lim_{x\to\infty} | z'(x) | = \infty . \quad \square$$

The following theorem gives a condition for the converse of Theorem 2.47 to hold.

THEOREM 2.48. If $A(x) \leqq 0$, $A'(x) - b(x) \leqq 0$, $b(x) \geqq 0$ for $x \in (a, \infty)$ and $\int_a^\infty t^4 b(t)\, dt = \infty$, then a necessary and sufficient condition for oscillatoricity of the differential equation (b) in (a, ∞) is that for each non-trivial, non-oscillatory z there should exist $c \in (a, \infty)$ with $z(x)z'(x)z''(x) \neq 0$ and

$$\operatorname{sgn} z(x) = \operatorname{sgn} z'(x) = \operatorname{sgn} z''(x) \text{ for } x \geqq c \tag{2.65}$$

while

$$\lim_{x\to\infty} | z(x) | = \lim_{x\to\infty} | z'(x) | = \infty . \tag{2.66}$$

PROOF. Necessity follows from Theorem 2.43. To prove sufficiency, we shall make use of the identity (1.8). For any solution z of the differential equation (b) we evidently have

$$-2F(z(x)) = z'^2(x) - 2z(x)z''(x) - 2Az^2(x) =$$
$$= -2F(z(\alpha)) - \int_\alpha^x b(t)z^2(t)\, dt . \tag{2.67}$$

Suppose that (2.65) and (2.66) hold true for any non-oscillatory solution z of the differential equation (b). With no loss of generality one may assume that $z(x) > 0$, $z'(x) > 0$, $z''(x) > 0$ and $z'''(x) \geqq 0$ for $x > c$. Then, clearly,

$$z(x) > \frac{z''(c)}{2}(x-c)^2$$

for $x > c$ and (2.57) implies that

$$-2F(z(x)) \leqq -2F(z(c)) - z''(c) \int_\alpha^x \frac{(t-c)^4}{2} b(t)\, dt$$

for $x > c$. It follows from the hypothesis that $\lim_{x\to\infty} F(z(x)) = \infty$. To prove the existence of an oscillatory solution, it is sufficient to show that there exists a non-trivial solution u of the differential equation (b) with $\lim_{x\to\infty} F(u(x))\neq\infty$. Suppose therefore, that z_1, z_2, z_3 is a fundamental set of solutions of (b) and let $\{u_n(x)\}$ be a sequence of solutions of (b) such that $u_n(n)=u_n'(n)=0$, $u_n''(n)\neq 0$, where n is a positive integer with $n>\alpha$. Let $u_n(x)=c_{1n}z_1(x) + c_{2n}z_2(x)+c_{3n}z_3(x)$, where $c_{1n}^2+c_{2n}^2+c_{3n}^2=1$. Using the same argument as in Theorem 1.13, one can show that, for some sequence $\{n_i\}$, the sequences $\{u_{n_i}(x)\}$, $\{u_{n_i}'(x)\}$, $\{u_{n_i}''(x)\}$ converge uniformly in every closed finite subinterval of (a, ∞) to functions $u(x)$, $u'(x)$, $u''(x)$, where u is a non-trivial solution of the differential equation (b). As $b(x)\geqq 0$, it follows from (2.67) that $-F(u_{n_i}(x))$ is a non-increasing function of x. On the other hand, since $F(u_{n_i}(n_i))=0$, we have $-F(u_{n_i}(x))\geqq 0$ for $\alpha\leqq x<n_i$. Letting n_i tend to infinity, we evidently get $-F(u(x))\geqq 0$ for $\alpha\leqq x<\infty$, hence $\lim_{x\to\infty} F(u(x))\neq\infty$. Thus, $u(x)$ is an oscillatory solution of the differential equation (b). The theorem is now proved. □

Let us return to the question whether the zeros of any two independent oscillatory solutions of the differential equation (b) separate each other.

LEMMA 2.15. If $A(x)\leqq 0$ and $b(x)\geqq 0$ for $x\in(a, \infty)$, then the derivative of any oscillatory solution of the differential equation (b) is bounded in $\langle\alpha, \infty)$, $a<\alpha<\infty$.

PROOF. Let u be any oscillatory solution of the differential equation (b) and let $\beta>\alpha$ be a zero of the second derivative of $u(x)$. Since the function

$$-2F(u(x))=u'^2(x)-2u(x)u''(x)-2A(x)u^2(x)=$$

$$=-2F(u(\alpha))-\int_\alpha^x b(t)u^2(t)\,dt$$

is non-increasing, we see that

$$u'^2(\beta)\leqq u'^2(\beta)-2A(\beta)u^2(\beta)=-2F(u(\beta))\leqq -2F(u(\alpha)).$$

The derivative $u'(x)$, in view of the oscillatoricity of $u(x)$, is then bounded for every $x\geqq\alpha$. Thus the lemma is proved. □

THEOREM 2.49. If $A(x) \leqq 0$, $A'(x) - b(x) \leqq 0$, $b(x) \geqq 0$ for $x \in (a, \infty)$, then the zeros of any two independent oscillatory solutions of the differential equation (b) separate each other in (a, ∞).

PROOF. It suffices to prove that the wronskian $W(u, v)$ of any two independent solutions u and v of the differential equation (b) has no zero in (a, ∞). Suppose the contrary, namely, let $W(u(\beta), v(\beta)) = 0$, $a < \beta < \infty$. Then there must exist non-zero constants c_1 and c_2 such that

$$c_1 u(\beta) + c_2 v(\beta) = 0 ,$$
$$c_1 u'(\beta) + c_2 v'(\beta) = 0 ,$$
$$c_1 u''(\beta) + c_2 v''(\beta) > 0 .$$

By Lemma 2.13, the solution $z = c_1 u + c_2 v$ has the property $\lim_{x\to\infty} z(x) = \lim_{x\to\infty} z'(x) = \infty$. On the other hand, the solutions u and v are oscillatory and hence, by Lemma 2.15, $u'(x)$ and $v'(x)$ are bounded in every interval $\langle \alpha, \infty)$, $a < \alpha < \infty$, hence $z'(x)$ is bounded in $\langle \alpha, \infty)$ as well. This is a contradiction, proving that $W(u(x), v(x)) \neq 0$ for $a < x < \infty$. The theorem is thus proved. □

REMARK 2.21. In Theorems 2.47, 2.48 and 2.49 we did not assume the property (v) for the function $b(x)$, but merely that $b(x) \geqq 0$ in (a, ∞). Yet the differential equation (b) is in class II. What is essential is that the second order differential equation (2.45) is disconjugate in (a, ∞), while $A'(x) - b(x) \leqq 0$ for $x \in (a, \infty)$.

From §1, Section 2 we know that if $b(x) \equiv 0$ in (a, ∞), then the differential equation (b) reduces to the self-adjoint equation (1.3) and that a necessary and sufficient condition for the oscillatoricity of (1.3) in (a, ∞) is the oscillatoricity of (1.4), i.e. of the equation

$$y'' + \frac{A}{2} y = 0 .$$

In the following theorem (Jones [74]) we show that a relationship exists between the oscillatoricity of (1.4) and of (b) or (a).

THEOREM 2.50. Let $b(x)$ be a function having the property (v) in

(a, ∞). Then the differential equation (b) is oscillatory in (a, ∞) whenever the differential equation (1.4) is oscillatory in (a, ∞).

PROOF. Let Y be a solution of the differential equation (a) with no zeros in (a, ∞) and let $Y(x)>0$ for $x \in (a, \infty)$. By Theorem 1.13, such a solution evidently exists and satisfies the identity (1.13) with $b=\infty$. Therefore,

$$Y'^2(x) - 2Y(x)Y''(x) - 2AY^2(x) \leqq 0 \quad \text{for } x \in (a, \infty).$$

Hence we get the following inequality:

$$\frac{2Y''(x)}{Y(x)} - \frac{Y'^2(x)}{Y^2(x)} \geqq -2A(x). \tag{2.68}$$

Let z_1, z_2 be independent solutions of the differential equation (b) with $Y(x) = z_1 z_2' - z_1' z_2$. The existence of such solutions can be proved as in Lemma 1.1. The band of solutions $u = c_1 z_1 + c_2 z_2$ satisfies the following equation of the form (c)

$$Yz'' - Y'z' + (Y'' + 2AY)z = 0,$$

which by the substitution $z = \sqrt{Y} v$ (cf. equation (2.13)) reduces to

$$v'' + \left[2A + \frac{3}{4}\left(\frac{2Y''}{Y} - \frac{Y'^2}{Y^2}\right)\right] v = 0. \tag{2.69}$$

It follows from (2.68) that

$$2A + \frac{3}{4}\left(\frac{2Y''}{Y} - \frac{Y'^2}{Y^2}\right) \geqq 2A - \frac{3}{2} A = \frac{A}{2}.$$

Now, if we apply the Sturm comparison theorem for second order equations (Sansone [124]), comparing equation (1.4) with (2.69), we see that whenever (1.4) oscillates, so does (2.69), and therefore the differential equation (b) is also oscillatory in (a, ∞). □

THEOREM 2.51. Let $A(x) \geqq 0$, $A'(x) - b(x) \leqq 0$, let $b(x)$ have the property (v) in (a, ∞) and let $\int_\alpha^\infty b(t)\,dt = \infty$, $a < \alpha < \infty$. Then the differential equation (b) is oscillatory in (a, ∞).

PROOF. Again, we use the integral identity (1.8), which may be written, for any solution z of the differential equation (b), in the form

$$-2F(z(x))=z'^2(x)-2z(x)z''(x)-2A(x)z^2(x)=$$
$$=-2F(z(\alpha))-2\int_\alpha^x b(t)z^2(t)\,dt\,.$$

Obviously, $-2F(z(x))$ is a decreasing function of x in $\langle\alpha,\infty)$, $a<\alpha<\infty$. Suppose that the differential equation (b) is non-oscillatory in (a,∞). Let z_1, z_2, z_3 be a fundamental set of solutions of the differential equation (b). Assume that z_n is a solution of (b) with $z_n(n)=z_n'(n)=0$, while $z_n=c_{1n}z_1+c_{2n}z_2+c_{3n}z_3$, where $c_{1n}^2+c_{2n}^2+c_{3n}^2=1$ and n is a positive integer greater than α. Assume, with no loss of generality, that $\lim_{n\to\infty} c_{in}=c_i$, $i=1,2,3$. Let $Z=c_1z_1+c_2z_2+c_3z_3$. Since $c_1^2+c_2^2+c_3^2=1$, Z is a non-trivial solution of the differential equation (b). Evidently, $-F(Z(x))>0$ for $x>\alpha$. Since $F(-Z(x))=F(Z(x))$ and the differential equation (b) is non-oscillatory in (a,∞), we may assume without loss of generality that $Z(x)>0$ for $x>\alpha$. Suppose that $Z'(\beta)=0$, $\beta>\alpha$. Then, at β, $Z(\beta)Z'(\beta)+A(\beta)Z^2(\beta)=F(Z(\beta))<0$. Hence $Z''(\beta)<0$ and Z' cannot have more than one zero strictly to the right of α. If $Z'(x)<0$ beginning with some x, then $Z'''(x)\geqq 0$, and equality holds at isolated points only. In this case, that sign of $Z(x)$ remains constant for sufficiently great x. If $Z''(x)$ is positive, a contradiction arises with $Z'(x)<0$. If $Z''(x)<0$ for large x, then in view of $Z'(x)<0$ we deduce that Z is negative, which again is a contradiction. Thus, $Z'(x)$ must be positive for large x. If we still have $Z(c)>0$ and $Z'(x)>0$ for $x\in\langle c,\infty)$, then

$$0<-2F(Z(x))=-2F(Z(c))-2\int_c^x b(t)Z^2(t)\,dt\leqq$$
$$\leqq -2F(Z(c))-Z^2(c)2\int_c^x b(t)\,dt\to-\infty \text{ as } x\to\infty\,.$$

The differential equation (b) must therefore have an oscillatory solution, and the theorem is proved. □

10. The Number of Oscillatory Solutions in a Fundamental System of Solutions of the Differential Equation (a)

Utz [140] raised the question, how many oscillatory solutions can be contained in a fundamental system of solutions of a given third order differential equation. His results were later generalized by Jones [69, 70].

An example has been given by Utz of a third order linear differential equation which may have a fundamental set of solutions with 0, 1, 2 or 3 oscillatory solutions in (a, ∞). It is the following equation with constant coefficients,

$$y''' - 3y'' + 4y' - 2y = 0 ,$$

for which a fundamental system of solutions is $y_1 = \exp(x) \sin x$, $y_2 = \exp(x) \cos x$, $y_3 = \exp(x)$. Reducing the above equation to the form (a), we get

$$y''' + y' = 0 ,$$

which is a self-adjoint equation. It is therefore natural to investigate this question for the differential equation (1.3), i.e. the self-adjoint equation

$$y''' + 2Ay' + A'y = 0 .$$

THEOREM 2.52. **A fundamental system for the differential equation (1.3) can be chosen to contain exactly 0, 1, 2 or 3 non-oscillatory solutions in (a, b).**

PROOF. It follows from Section 2 in §1 that the general solution of the differential equation (1.3) can be written in the form

$$y = c_1 y_1^2 + c_2 y_1 y_2 + c_3 y_2^2 ,$$

where y_1, y_2 is a fundamental system of solutions of the second order differential equation (1.4) and c_1, c_2, c_3 are arbitrary constants. It is well known (Sansone [124]) that the zeros of y_1 and y_2 (if they exist) separate each other and hence if one solution oscillates in (a, b), so do all the solutions of the equation (1.4).

In view of this fact and under the assumption that the differential equation (1.4) is oscillatory in (a, b), there exist the following four fundamental systems of solutions of the differential equation (1.3):

$$\begin{array}{lll} y_1(x)y_2(x), & y_1(x)\,(y_1(x)+y_2(x)), & y_2(x)\,(y_1(x)+y_2(x)), \\ y_1(x)y_2(x), & y_1(x)\,(y_1(x)+y_2(x)), & y_1^2(x)+y_2^2(x), \\ y_1(x)y_2(x), & y_1^2(x)+2y_2^2(x), & 2y_1^2(x)+y_2^2(x), \\ y_1^2(x)+2y_2^2(x), & 2y_1^2(x)+y_2^2(x), & (y_1(x)+y_2(x))^2+y_1^2(x). \end{array}$$

It remains to prove that if the equation (1.3) has an oscillatory solution, then the equation (1.4) is oscillatory in (a, b).

Suppose that y_1 is a solution of the differential equation (1.4) with $y_1(\alpha)=0$, $y_1'(\alpha)\neq 0$, $a<\alpha<b$. By Theorem 1.14, whenever one solution of (1.3) oscillates in (a, b), so do all those solutions which have a zero. Thus, $y_1^2(x)$ oscillates in (a, b), and the theorem is proved. □

In the sequel, we shall investigate the differential equation (a) from class I or from class II, respectively.

LEMMA 2.16. If the differential equation (a) is in class I in (a, b) and has an oscillatory solution in (a, b), then there exists a non-trivial, non-oscillatory solution $y(x)$ of (a) with $y(x)>0$ for $x\in(a, b)$.

The proof, by construction, is analogous to that of Theorem 1.13. The solution constructed in that way satisfied $y(x)\geqq 0$ for $x\in(a, b)$, because $y_n(x)\geqq 0$ for $x\in(a, x_n)$. On the other hand, it is impossible that $y(\alpha)=0$, $a<\alpha<b$, otherwise y would oscillate in (a, b). Therefore, $y(x)\neq 0$ in (a, b). □

THEOREM 2.53. If the differential equation (a) is in class I and is oscillatory in (a, b), then there exists a fundamental system of solutions of (a) with three oscillatory solutions and there is a fundamental system for (a) with two solutions oscillatory in (a, b).

PROOF. Let $\alpha<\beta\in(a, b)$. Let y_1, y_2, y_3 be a fundamental set of solutions such that $y_1(\alpha)=y_1'(\alpha)=0$, $y_1''(\alpha)\neq 0$, $y_2(\alpha)=y_2''(\alpha)=0$, $y_2'(\alpha)\neq 0$, $y_3(\beta)=y_3'(\beta)=0$, $y_3''(\beta)\neq 0$. The solutions y_1, y_2 belong to the band at α, hence they are oscillatory in (α, b). All the solutions constructed as linear combinations of these two have simple zeros for $x>\alpha$. Therefore, y_3 is independent of y_1 and y_2. The solution y_3, being in the band at β, oscillates also.

To prove the second part of the theorem, it is sufficient to choose a fundamental system y_1, y_2, z_3 in which y_1, y_2 have the above properties and z_3 is a solution without zeros in (a, b), whose existence is guaranteed by Lemma 2.16. □

THEOREM 2.54. If the differential equation (b) is in class II and some solution of (b) oscillates in (a, b), then there exists a fundamental system of solutions which has i oscillatory solutions, $i=0, 1, 2$.

PROOF. Since the differential equation (b) has an oscillatory solution, its adjoint differential equation (a) oscillates in (a, b) and is in class I in (a, b). Let u_1, u_2, u_3 be a fundamental system of solutions of (a) such that $u_1 \neq 0$ in (a, b) and u_2, u_3 are oscillatory in (a, b). Also, u_1, u_2, w_3, where $w_3(x)=u_3(x)+\lambda u_1(x)$, is a fundamental system of solutions of (a), λ being chosen so that $u_3(\alpha)+\lambda u_1(\alpha)=0=u_2(\alpha)$, $a<\alpha<b$. The solution w_3 of (a) is oscillatory also.

The functions z_1, z_2, z_3, where $z_1=u_1u_2'-u_1'u_2$, $z_2=u_1w_3'-w_3u_1'$, $z_3=u_2w_3'-u_2'w_3$, form a fundamental set of solutions of the differential equation (b) (cf. §1, Section 2). Evidently, z_1, z_2 are oscillatory solutions, because u_1 is non-oscillatory in (a, b) and it follows from the properties of bands that if the wronskian of u_1, u_2 (that is, the solution z_1) were different from zero in some (α, b), then u_1 would necessarily oscillate. The situation is similar with z_2. The solution z_3 satisfies $z_3(\alpha)=z_3'(\alpha)=0$, hence $z_3\neq 0$ for $x>\alpha$, and therefore z_3 is a non-oscillatory solution, since (b) is in class II.

Let u_1, u_2, u_3 be a fundamental set of solutions of the differential equation (a) such that u_1, u_2, u_3 are oscillatory in (a, b). Let $u_1(\alpha)=0$, $a<\alpha<b$, and let $u_3(\alpha)\neq 0$. Choose a constant λ so that $u_2(\alpha)+\lambda u_3(\alpha)=0$ and let $v_2=u_2+\lambda u_3$. The functions u_1, v_2, u_3 form a fundamental set of solutions of the differential equation (a), all of them being oscillatory in (a, b). Since $u_1(\alpha)=v_2(\alpha)=0$, the solutions u_1 and v_2 belong to the band at α, and hence their zeros separate each other in (α, b). Let $\beta>\alpha$ be the first zero of u_1 to the right of α and let

β_1 be the first zero of v_2 to the right of α. Let $\beta<\beta_1$. Since $u_1(\beta)v_2(\beta_1)-u_1(\beta_1)v_2(\beta)\neq 0$, the system of equations

$$c_1u_1(\beta)+c_2v_2(\beta)+c_3u_3(\beta)=0\,,$$
$$c_1u_1(\beta_1)+c_2v_2(\beta_1)+c_3u_3(\beta_1)=0,\ \ c_3\neq 0\,,$$

can be solved with respect to c_1, c_2. With constants c_1, c_2 found in this way, we construct the solution $v_3=c_1u_1+c_3u_3$. As $c_1\neq 0$, the functions u_1, v_2, v_3 form a fundamental system of solutions of the differential equation (a) and all of them are oscillatory in (a, b). The functions z_1, z_2, z_3, where $z_1=u_1v_2'-u_1'v_2$, $z_2=u_1v_3'-v_3u_1'$, $z_3=v_2v_3'-v_2'v_3$, evidently form a fundamental system of solutions of the differential equation (b). Clearly, $z_1(\alpha)=z_1'(\alpha)=0$, $z_2(\beta)=z_2'(\beta)=0$, $z_3(\beta_1)=z_3'(\beta_1)=0$, so z_1, z_2, z_3 are non-oscillatory solutions of the equation (b), which is in class II. The fact that (b) has a fundamental system with exactly one oscillatory solution is obvious for it is sufficient to choose an oscillatory solution for z_1 and to take z_2, z_3 having double zeros at $\alpha<\beta\in(a, b)$. Thus, the theorem is proved. □

THEOREM 2.55. If the differential equation (b) is in class II in (a, b) and has a fundamental system consisting of three oscillatory solutions, then there exists a fundamental set of solutions of the differential equation (a) with one oscillatory and two non-oscillatory solutions.

PROOF. Let u_1, u_2, u_3 be a fundamental set of solutions of the differential equation (a) with $u_1(\alpha)=u_2(\alpha)=0$, $a<\alpha<b$, $u_3(x)>0$ for $x\in(a, b)$. The functions $z_1=u_1u_3'-u_1'u_3$, $z_2=u_2u_3'-u_2'u_3$, $z_3=u_1u_2'-u_1'u_2$ form a fundamental system of solutions of the differential equation (b) such that z_1, z_2 are oscillatory in (a, b) and z_3 is non-oscillatory in (a, b). Since $z_1z_2'-z_1'z_2=ku_3\neq 0$ for $x\in(a, b)$, $k\neq 0$, the zeros of z_1 and z_2 separate each other. If the differential equation (b) has a fundamental system with three oscillatory solutions, there exists a solution z of (b) of the form $z=z_3+c_1z_1+c_2z_2$. Let $a<x_1<x_2<\ldots$ be the zeros of the solution z in (a, b). Define the sequence

$$y_n(x)=k_{1n}z_1(x)+k_{2n}z_2(x)\,,$$

where $k_{1n}^2+k_{2n}^2=1$ and $y_n(x_n)=0$. The zeros of $y_n(x)$ and $z(x)$ to the left of x_n separate each other (bands the adjoint equation). Without loss of generality we may assume that $\lim_{n\to\infty} k_{in} = k_i$ for $i=1, 2$. Let $y(x) = k_1z_1(x) + k_2z_2(x)$. Since $\{y_n(x)\}$ converges to $y(x)$ uniformly on $\langle x_j, x_{j+1}\rangle$ and since each $y_n(x)$ for $n>j+2$ has a zero in (x_j, x_{j+1}), then necessarily y also has a zero in (x_j, x_{j+1}). Since $k_1^2+k_2^2=1$, the solutions y and z are linearly independent, so that y and z cannot have two zeros in common. It follows that for $j\geqq N$, where $N>0$ is sufficiently great, the solution y has a zero in (x_j, x_{j+1}). Since y is a solution of the equation (b) which is in class II, it must change its sign in the interval (x_j, x_{j+1}). Suppose that $y(x_0)z'(x_0)-y'(x_0)z(x_0)=0$, $a<x_0<b$. The algebraic equations

$$l_1y(x_0)+l_2z(x_0)=0\,, \quad l_1y'(x_0)+l_2z'(x_0)=0$$

have a non-trivial solution for l_1 and l_2. Let

$$w(x)=l_1y(x)+l_2z(x)\,.$$

From $w(x_0)=w'(x_0)=0$ it follows that $w(x)\neq 0$ for $x>x_0$. This is impossible, however, because for $j\geqq N$ and $x_j>x_0$ such that $l_2z(x)\geqq 0$ in $\langle x_j, x_{j+1}\rangle$ there necessarily exists a point $\alpha_1\in(x_j, x_{j+1})$ with $l_1y(\alpha_1)\geqq 0$ and a point $\beta_1\in(x_{j+1}, x_{j+2})$ with $l_1y(\beta)\leqq 0$. Thus $l_1y(\alpha_1)+l_2z(\alpha_1)\geqq 0$ and $l_1y(\beta_1)+l_2z(\beta_1)\leqq 0$. Therefore, $y(x)z'(x)-z(x)y'(x)$ is a non-oscillatory solution of the differential equation (a), since it does not vanish at x_0. Evidently, $yz'-zy' = (k_1z_1+k_2z_2)(z_3'+c_1z_1'+c_2z_2') - (z_3+c_1z_1+c_2z_2)(k_1z_1'+k_2z_2') = k(k_1u_1+k_1c_2u_3+k_2u_2+k_2c_1u_3)$, where $k\neq 0$. Since k_1 and k_2 are not both equal to zero, the functions $yz'-y'z$ and u_3 are two linearly independent solutions of the differential equation (a), which completes the proof. □

THEOREM 2.56. If the differential equation (a) is in class I in (a, b) and has at least one oscillatory solution in (a, b) and if it has a fundamental set of solutions with two or three non-oscillatory solutions, then the differential equation (b) has a fundamental system consisting of three oscillatory solutions.

PROOF. If (a) has a fundamental system of solutions with three oscillatory solutions, evidently there exists a fundamental system for (a) with exactly one oscillatory solution. Let u_1, u_2, u_3 be a fundamental system of solutions of (a), where u_1 is oscillatory and u_2 and u_3 are non-oscillatory. Let $u_2(\alpha)=u_3(\alpha)>0$, $a<\alpha<b$. The functions $w_1=u_1u_2'=u_1'u_2$, $w_2=u_1u_3'-u_1'u_3$, $w_3=u_2u_3'-u_2'u_3$ form a fundamental set of solutions of the differential equation (b). Observe that w_1, w_2 are oscillatory in (a, b). Since $u_2(\alpha)=u_3(\alpha)$, $y=u_2-u_3$ is an oscillatory solution of the differential equation (a). Let $\alpha<\alpha_1<\alpha_2<\ldots$ be consecutive zeros of the solution y. Then $y'(\alpha_i)y'(\alpha_{i+1})<0$. Clearly, $w_3(\alpha_i) = u_2(\alpha_i)u_3'(\alpha_i) - u_2'(\alpha_i)u_3(\alpha_i) = u_2(\alpha_i)(u_3(\alpha_i)-u_2(\alpha_i))$ has opposite signs at each two consecutive zeros α_i of the solution y. Thus the theorem is proved. □

As an example, consider the differential equation

$$y'''+y'+\left(\frac{2}{e^x+2}\right)(y+y'')=0\,. \tag{2.70}$$

Its general solution is $y=c_1\sin x+c_2\cos x+c_3(1+e^{-x})$. Evidently, $\sin x$, $2(1+e^{-x})+\cos x$, $1+e^{-x}$ form a fundamental set of solutions of (2.70) with one oscillatory and two non-oscillatory solutions. Transforming the equation (2.70) to the normal form (a) by the substitution $y=z\exp\left(-\frac{1}{3}\int_a^x P(t)\,dt\right)$, where $P(t)=2/(e^t+2)$, we get and equation which has the same oscillation properties as (2.70) and is in class I whenever (2.70) is in class I.

The differential equation (2.70) is in class I, otherwise it would necessarily have a non-trivial solution y with $y(\alpha-\delta)=y(\alpha)=y'(\alpha)=0$ for $a<\alpha<b$ and $\delta>0$. But this is impossible, since

$$\begin{vmatrix} \sin\alpha, & \cos\alpha, & 1+e^{-\alpha} \\ \cos\alpha, & -\sin\alpha, & -e^{-\alpha} \\ \sin(\alpha-\delta), & \cos(\alpha-\delta), & 1+e^{-\alpha+\delta} \end{vmatrix} =$$

$$=e^{-\alpha}[\cos\delta+\sin\delta-e^{\delta}]+\cos\delta-$$

$$-1<e^{-\alpha}[\cos\delta+\sin\delta-1-\delta]\leq 0\,.$$

Applying Theorem 2.56, we conclude that the adjoint equation to (2.70) satisfies the conditions of Theorem 2.55.

11. Criteria for Oscillatoricity of Solutions of the Differential Equation (a) in the Case that the Laguerre Invariant Does Not Satisfy Condition (v)

The results given in this section have been derived under the assumption that the function $b(x) \geqq 0$ for $x \in (a, b)$ (Rovder [121, 123]). Thus, in a sense, they generalize some of the results from the previous sections. However, under the above assumption, the separation property of zeros for bands of solutions of the differential equation (a) is not preserved in general.

Lemma 2.17. Let u and v be functions having derivatives of the second order in (a, b). Further, let $u(\alpha) = u'(\alpha) = 0$, $u''(\alpha) > 0$, $u(\beta) = 0$, $u(x) > 0$ in (α, β), where $a < \alpha < \beta < b$. Let the function v satisfy $v(\alpha) \geqq 0$, $v'(\alpha) > 0$, $v(x) > 0$ in $(\alpha, \beta\rangle$. Then there exist a number $\tau \in (\alpha, \beta)$ and a constant $c > 0$ such that the function $f(x) = v(x) - cu(x)$ has a double zero at τ and also $f''(\tau) \geqq 0$ and $f(x) \geqq 0$ in $\langle \alpha, \tau\rangle$.

Proof. Define the function

$$w(x) \overset{\text{d}}{=} \begin{cases} \dfrac{u(x)}{v(x)} & \text{for } x \in (\alpha, \beta> , \\ 0 & \text{for } x = \alpha . \end{cases}$$

Evidently, w is continuous in $\langle \alpha, \beta\rangle$, while $w(\alpha) = w(\beta) = 0$ and $w(x) > 0$ in (α, β). Therefore, w attains its maximum over $\langle \alpha, \beta\rangle$ at some $\tau \in (\alpha, \beta)$. Denote this maximum by $1/c$, where c is a positive number. Then $w(x) \leqq 1/c$ and $w(\tau) = 1/c$, for $x \in (\alpha, \beta\rangle$. Thus,

$$\frac{u(x)}{v(x)} \leqq \frac{1}{c}, \quad \frac{u(\tau)}{v(\tau)} = \frac{1}{c}.$$

These inequalities, together with $v(\alpha) \geqq 0$ and $u(\alpha) = 0$, imply that $v(x) - cu(x) \geqq 0$ for $x \in \langle \alpha, \beta\rangle$, $v(\tau) - cu(\tau) = 0$, $\alpha < \tau < \beta$. The function $f(x) = v(x) - cu(x)$ therefore attains its minimum over $\langle \alpha, \beta\rangle$ at τ and $f(x) \geqq 0$ for all $x \in \langle \alpha, \beta\rangle$. Since the functions u, v have derivatives of the second order in $\langle \alpha, \beta \rangle$, obviously $f'(\tau)=0$. Thus the lemma is proved. □

Alongside the differential equation (a), we shall now investigate also the differential equation (a_1). We prove the following comparison theorem.

THEOREM 2.57. Let the coefficients of the differential equations (a) and (a_1) satisfy

$$A_1(x) \geqq A(x),\ b_1(x) \geqq b(x),\ b_1(x) \geqq 0,\ x \in (a, b). \qquad (2.71)$$

Let $\alpha < \beta \in (a, b)$ be two consecutive zeros of a solution y of the differential equation (a) and let α be a double zero of y. Then any solution z of the differential equation (a_1) having a simple zero at α has another zero in $(\alpha, \beta\rangle$.

PROOF. Let $\bar{y}$ be a solution of (a) such that $\bar{y}(\alpha)=\bar{y}(\alpha)=0$, $\bar{y}(\beta)=0$. Suppose that $\bar{y}''(\alpha)>0$ and $\bar{y}(x)>0$ in (α, β). Let z be a solution of (a_1) with $z(\alpha)=0$, $z'(\alpha)>0$. Suppose that z has no zero in $(\alpha, \beta\rangle$, but $z(x)>0$ in $(\alpha, \beta\rangle$. By Lemma 2.17, there is a constant $c>0$ such that the function $\bar{f}(x) \overset{d}{=} z(x)-c\bar{y}(x)$ has a double zero at $\tau \in (\alpha, \beta)$ and, at the same time, $\bar{f}(x) \geqq 0$ for $x \in <\alpha, \tau>$ and $\bar{f}(\tau) \geqq 0$. Denoting $c\bar{y}$ by y, we see that the function $f(x)=z(x)-y(x)$ satisfies

$$f(\tau)=f'(\tau)=0,\ f''(\tau) \geqq 0,\ f(x) \geqq 0,\ x \in \langle \alpha, \tau \rangle\,. \qquad (2.72)$$

Multiplying (a_1) by z and (a) by y, we get

$$\left[zz''-\frac{1}{2}z'^2+A_1z^2\right]'=-b_1z^2\,,$$

$$\left[yy''-\frac{1}{2}y'^2+Ay^2\right]'=-by^2\,.$$

Subtraction and integration from α to τ of the above identities yields

$$\left[\left(yy''-\frac{1}{2}y'^2+Ay^2\right)-\left(zz''-\frac{1}{2}z'^2+A_1z^2\right)\right]_\alpha^\tau=$$

$$=\int_\alpha^\tau [b_1z^2-by^2]\,dt\,. \tag{2.73}$$

Since $f(x)\geqq 0$ in $\langle\alpha,\tau\rangle$, we have also $z(x)\geqq y(x)\geqq 0$ there. Then $z^2(x)\geqq y^2(x)$ for $x\in\langle\alpha,\tau\rangle$. This inequality and (2.71) imply that $b_1(x)z^2\geqq b(x)y^2(x)$ for $x\in\langle\alpha,\tau\rangle$. Thus the right-hand side of (2.73) is non-negative, and hence

$$\left[\left(yy''-\frac{1}{2}y'^2+Ay^2\right)-\left(zz''-\frac{1}{2}z'^2+A_1z^2\right)\right]_\alpha^\tau\geqq 0\,.$$

As $y(x)=c\bar{y}(x)$, we have $y(\alpha)=y'(\alpha)=0$, $y''(\alpha)>0$ and, further, $z(\alpha)=0$, $z(\tau)=y(\tau)$ and $z'(\tau)=y'(\tau)$. Substituting from these relations in the above inequality, we obtain

$$-y(\tau)\,[z''(\tau)-y''(\tau)]-\frac{1}{2}[y'^2(\tau)-z'^2(\tau)]-$$

$$-y^2(\tau)\,[A_1(\tau)-A(\tau)]-\frac{1}{2}z'^2(\alpha)\geqq 0\,,$$

that is,

$$-y(\tau)f''(\tau)-y'(\tau)f'(\tau)-y^2(\tau)\,[A_1(\tau)-A(\tau)]-$$

$$-\frac{1}{2}z'^2(\alpha)\geqq 0\,.$$

From (2.71) and (2.72) it follows that the left-hand side of the last inequality is negative, giving a contradiction. Therefore, the solution z of the differential equation (a_1) must have a zero in (α,β). □

COROLLARY 2.13. Assume (2.71) and let the differential equation (a) be in class I and oscillatory in (a, b). The differential equation (a_1) is also oscillatory in (a, b).

The proof is obvious, because from the properties of bands it follows that every solution of (a) having a double zero has another, simple, zero to the right of the double one. Theorem 2.57 then implies the assertion of this theorem. □

THEOREM 2.58. Let the coefficients of the differential equations (a) and (a_1) satisfy

$$A_1(x) \geqq A(x), \quad b_1(x) \geqq b(x) \geqq 0, \quad x \in (a, b) . \tag{2.74}$$

If the differential equation (a) is oscillatory in (a, b), then so is also the differential equation (a_1).

PROOF. If (a) is oscillatory in (a, b) and $b(x) \geqq 0$ for $x \in (a, b)$, then (a_1) is oscillatory by Theorem 2.57. □

THEOREM 2.59. Let $b(x) \geqq 0$ for $x \in (a, b)$. The differential equation (a) is oscillatory in (a, b) if and only if its adjoint equation (b) is oscillatory in (a, b).

PROOF. Necessity. Let the differential equation (a) be oscillatory in (a, b) and let $u(x)$ be its solution without zeros in (a, b). Such a solution exists by Theorem 1.15. Let $v(x)$ be an oscillatory solution of (a). The function $w(x) = u(x)v'(x) - u'(x)v(x)$ satisfies the adjoint equation (b). If $x_1 < x_2 \in (a, b)$ be neighbouring zeros of v, then

$$\int_{x_1}^{x_2} \frac{u(t)v'(t) - u'(t)v(t)}{u^2(t)} \, dt = \left[\frac{v(t)}{u(t)}\right]_{x_1}^{x_2} = 0 .$$

Therefore, w has a zero between every two neighbouring zeros of v, and hence the equation (b) is oscillatory in (a, b).

Sufficiency. Let the differential equation (b) be oscillatory in (a, b). Let z be a solution of (b) such that $z(\alpha) = z'(\alpha) = 0$, $z'(\alpha) = 1$, $a < \alpha < b$. The integral identity (1.8) implies that z has no simple zeros to the right of α, therefore, $z(x) \geqq 0$ for $x > \alpha$. There are two possible cases: either $z(x)$ has infinitely many double zeros in (α, β), or there exists a number $\beta > \alpha \in (a, b)$ such that $z(x) > 0$ in (β, b). Let $z_1(x)$ be a solution of (b), linearly independent of $z(x)$. The function $w(x) = z_1(x)z'(x) - z(x)z_1'(x)$ satisfies (a) and is oscillatory in either case. In the first case, the equation (a) is oscillatory due to w vanishing at double zeros of t. In the second case, we assume that $z_1(x)$ is an oscillatory solution of (b) and continue as in the necessity part of the proof. □

COROLLARY 2.14. Suppose the inequalities (2.74) are true. If the differential equation (b) is oscillatory in (a, b), then so is the differential equation (b_1) adjoint to (a_1).

This assertion follows from Theorems 2.58 and 2.57. □

COROLLARY 2.15. Let $A_1(x) \geqq A(x)$, $b_1(x) \geqq b(x)$ and $b_1(x) \geqq 0$. Assume that the differential equation (b) belongs to class II and is oscillatory in (a, b). Then the differential equation (b_1) is also oscillatory in (a, b).

PROOF. It follows from the hypotheses that the coefficients of (a) and (a_1) satisfy the assumption (2.71) from Theorem 2.57. Since the equation (b) is in class II, the equation (a) belongs to class I and is oscillatory. By Corollary 2.13, the differential equation (a_1) is oscillatory in (a, b) and hence by Theorem 2.59, the equation (b_1) is oscillatory as well. □

THEOREM 2.60. Let $b(x) \geqq 0$ for $x \in (a, b)$ and let the second order differential equation

$$y'' + \frac{1}{2} A(x) y = 0$$

be oscillatory in (a, b). Then the differential equation (a) is oscillatory in (a, b).

The proof follows by comparison of (a) with the self-adjoint differential equation. □

Comparison theorems enable us to derive further criteria for oscillation of solutions of the differential equations (a) and (b).

To begin with, consider the following differential equation with constant coefficients

$$u''' + pu' = 0, \quad \text{where } p > 0 \text{ is a constant},$$

or the so-called Euler differential equation (2.43), which can be written in the form

$$u''' + \frac{p}{x^2} u' - \frac{p}{x^3} u = 0, \quad \text{where } p > 1 \text{ is a constant},$$

which is oscillatory. Here, the Laguerre invariant $b(x) \equiv 0$.

THEOREM 2.61. Let $b(x) \geqq 0$ and $2A(x) \geqq p$, $p>0$, or $2A(x) \geqq p/x^2$ for $p>1$, $x \in (a, \infty)$. Then both differential equations (a) and (b) are oscillatory in (a, ∞).

The proof follows from Theorem 2.58 and Corollary 2.14. It should be noted that Theorem 2.61 generalizes Theorem 2.36. □

Further oscillation criteria result from comparison of the differential equation (a) with the following equation with constant coefficients

$$y''' + pu' + qu = 0 , \tag{2.75}$$

where p, q are constants.

REMARK 2.22. A simple computation shows that the following results hold for the equation (2.75):

a) If $p<0$ and $q>0$, then the differential equation (2.75) is oscillatory in (a, ∞) if and only if

$$q - \frac{2}{3\sqrt{3}}(-p)^{3/2} > 0 .$$

Moreover, all the solutions of (2.75) are then oscillatory in (a, ∞), except one (up to linear dependence), which tends monotonically to zero together with its first and second derivatives as $x \to \infty$.

b) If $p<0$, then the differential equation (2.75) is oscillatory in (a, ∞) if and only if

$$-q - \frac{2}{3\sqrt{3}}(-p)^{3/2} > 0 .$$

It can be shown that the differential equation (2.75) has two independent oscillatory solutions whose zeros separate each other in (a, ∞). Moreover, the absolute values of consecutive maxima and minima (of oscillatory solutions) form a decreasing sequence.

We can easily see that the result b) can be derived by taking into consideration, in the case a), the equation adjoint to (2.75).

c) If $p>0$, $q>0$, then all the solutions of (2.75) are oscillatory in (a, ∞), except one solution (up to linear dependence), which together with its derivatives tends to zero as $x \to \infty$.

Besides the equation with constant coefficients, let us consider also the Euler differential equation in the form

$$u''' + \frac{p}{x^2} u' - \frac{\varepsilon - p}{x^3} u = 0 , \tag{2.76}$$

where $p<1$, $\varepsilon>0$, which is oscillatory if and only if

$$\varepsilon > \frac{2}{3\sqrt{3}} (1-p)^{3/2} .$$

THEOREM 2.62. Assume that either $2A(x)>p$ and $b(x) \geqq q$ for $x \in (a, \infty)$, where $p \leqq 0$ and $q > \frac{2}{3\sqrt{3}} (-p)^{3/2}$, p, q being constants, or

$$A(x) \geqq \frac{p}{2x^2}, \quad b(x) \geqq \frac{\varepsilon}{x^3} ,$$

where

$$p>1, \quad \varepsilon > \frac{2}{3\sqrt{3}} (1-p)^{3/2} ,$$

p, ε being constants, $x \in (0, \infty)$. Then the differential equation (a) is oscillatory.

The proof follows again by comparison of the equation (a) with (2.75) or (2.76). □

Owing to the fact that, in either case, the differential equation (a) is in class I, the differential equation (b) is also oscillatory.

The above result can be applied to the equation

$$y''' + (\sin x) y' + \frac{1}{2} \left(\cos x + \frac{2\sqrt{2}}{\sqrt{3}} \right) y = 0 ,$$

where $A(x) = \frac{1}{2} \sin x$, $b(x) = \frac{1}{2} \left(\cos x + \frac{2}{3} \sqrt{2} \right) - \frac{1}{2} \cos x = \frac{2}{3\sqrt{2}} > \frac{2}{3\sqrt{3}} (-p)^{3/2} > 0$, $A(x) \geqq -\frac{1}{2}$, that is, $2A(x) \geqq -1 = p$. The given equation is thus oscillatory in (a, ∞).

A criterion analogous to the above theorems can be established for non-oscillatoricity of the differential equation (a).

THEOREM 2.63. Let the differential equation (a) be in class I and let $b(x)\geqq 0$ for $x\in(a, \infty)$. Moreover, assume that the following conditions are fulfilled:

a) $2A(x)\leqq p$, $b(x)\leqq q$, where $p\leqq 0$, $q\leqq\frac{2}{3\sqrt{3}}(-p)^{3/2}$, p, q being constants, $x\in(a, \infty)$.

b) $2A(x)\leqq\frac{p}{x^2}$, $b(x)\leqq\frac{\varepsilon}{x^3}$, $p\leqq 1$, $\varepsilon\leqq\frac{2}{3\sqrt{3}}(1-p)^{3/2}$, p, ε being constants, $x\in(0, \infty)$.

Then the differential equation (a) is non-oscillatory in (a, ∞).

The proof uses again comparison of the equation (a) with (2.75) or (2.76), assuming that (a) is oscillatory in (a, ∞). From the hypotheses of the theorem and from Corollary 2.15 it follows that (2.75) or (2.76) would be oscillatory, which contradicts the assumption concerning the coefficients of (2.75) and (2.76). □

The following theorem deals with the same questions as the preceding Section 10.

THEOREM 2.64. Let $b(x)\geqq 0$ and $2A(x)\geqq p$ or $2A(x)\geqq px^{-2}$, $p>1$, $x\in(a, \infty)$ or $x\in(0, \infty)$. Then there is a fundamental system for the equation (a) which contains two oscillatory solutions and one non-oscillatory solution. The non-oscillatory solution has no zeros in (a, ∞) or $(0, \infty)$, respectively.

PROOF. Theorem 2.61 implies that the differential equation (a) is oscillatory in the given interval. From Theorem 1.14 it follows that every solution having a zero oscillates in the given interval and, by Theorem 1.15, there is at least one solution of (a) with no zeros in the given interval. These facts imply the assertion of this theorem. □

THEOREM 2.65. Let $b(x)\geqq 0$ for $x\in(a, \infty)$ and let the differential equation (a) be non-oscillatory in (a, ∞). Then there is a number $\gamma>a$ such that the differential equation (a) is disconjugate in $\langle\gamma, \infty)$.

PROOF. The differential equation (a) being non-oscillatory in (a, ∞), there exist a solution y of (a) and a number $\gamma>a$ such that $y(\gamma)=0$

and $y(x) \neq 0$ for $x > \gamma$. Let z be a solution of (a) with $z(\gamma) = z'(\gamma) = 0$, $z''(\gamma) \neq 0$. If $y'(\gamma) \neq 0$, then Theorem 2.57 implies that $z(x) \neq 0$ for $x > \gamma$. If $y'(\gamma) = 0$, then $z(x) = cy(x)$, $c \neq 0$ is a constant, and hence $z(x) \neq 0$ for $x > \gamma$.

We now show that no solution of (a) has more than two zeros in $\langle \gamma, \infty)$. Let us suppose that $u(\gamma) = u(x_1) = u(x_2) = 0$, $\gamma \leqq x_1 \leqq x_2$. If $\gamma = x_1 < x_2$, then $u(x) = cz(x)$, therefore $u(x) \neq 0$ for $x > \gamma$. Let $\gamma < x_1 = x_2$. This case is impossible due to the identity (1.6). Let $\gamma < x_1 < x_2$. Assume $u(x) > 0$ in the interval (x_1, x_2). By Lemma 1.2, there exist $c > 0$ and $\tau \in (x_1, x_2)$ such that the solution $z(x) - cu(x)$ has a double zero at τ and a simple zero at γ, which contradicts (1.6). Thus, no solution of the differential equation (a) having a zero at γ can have more than two zeros in $\langle \gamma, \infty)$.

Now let v be a solution of the differential equation (a) with $v(\gamma) \neq 0$ and let $v(x_1) = v(x_2) = v(x_3) = 0$, $\gamma < x_1 < x_2 < x_3$. Then, as in the previous case where $x_1 < x_2 = x_3$, a contradiction arises. Let $v(x) > 0$ in (x_2, x_3). Let w be a solution of the differential equation (a) such that $w(\gamma) = w(x_1) = 0$ and $w(x) < 0$ in (γ, x_1). We have $w(x) > 0$ in (x_1, ∞). By Lemma 2.16, for some $c > 0$ and some $\tau \in (x_1, x_2)$, the solution $w(x) - cv(x)$ has a double zero at τ and a simple zero at y_1, which contradicts the identity (1.6). Thus the theorem is proved. □

REMARK 2.23. Comparison theorems for three differential equations of the form (a) can be derived analogously as for two equations. Several results in this direction are established in Rovder's paper [2].

12. The Case, When the Laguerre Invariant Is an Oscillatory Function of x

In this section, we shall deduce sufficient conditions in order that, under the assumption that the function $b(x)$ is oscillatory in (a, b), the differential equation (a) should be disconjugate in (a, b) (Greguš [56]).

Let there be given a second order differential equation of the form

$$wy'' - w'y' + [w'' + 2Aw]\, y = 0\,, \tag{c}$$

where $w = w(x)$ is a given function, continuous together with its derivatives up to order three inclusive, and let $w(x) \neq 0$ in (a, b). Moreover, let $A = A(x)$ be a function with a continuous first derivative in (a, b).

If w is a solution of the differential equation (b), then by termwise differentiating the equation (c) we get the equation (a) with

$$b(x) = A'(x) + 2A(x)\frac{w'(x)}{w(x)} + \frac{w'''(x)}{w(x)}. \tag{2.77}$$

Let $w(x)$ and $b(x)$ be given functions of $x \in (a, b)$. In order to determine the coefficient $A(x)$ we get, from (2.77), the first order differential equation

$$A'(x) + 2\frac{w'(x)}{w(x)}A(x) = b(x) - \frac{w'''(x)}{w(x)}.$$

The last equation is satisfied by the function

$$A_0(x) = \frac{1}{w^2(x)}\left[A_0(x_0) + \int_{x_0}^{x} [b(t)w^2(t) - w'''(t)w(t)]\,\mathrm{d}t\right],$$

$$a < x_0 < b; \tag{2.78}$$

obviously,

$$A_0'(x) = -\frac{2w'(x)}{w(x)}A_0(x) + \frac{w(x)b(x) - w'''(x)}{w(x)}. \tag{2.79}$$

Let A_0, A_0' be defined by (2.78), (2.79). We shall consider the equations

$$y''' + 2A_0y' + [A_0' + b]\,y = 0, \tag{a$_0$}$$

$$z''' + 2A_0z' + [A_0' - b]\,z = 0, \tag{b$_0$}$$

$$wy'' - w'y' + [w'' + 2A_0w]\,y = 0. \tag{c$_0$}$$

REMARK 2.24. The function A_0 depends on $x \in (a, b)$ and on the parameter $A_0(x_0)$.

COROLLARY 2.16. A sufficient condition for the differential equation (a_0) not to be disconjugate in (a, b) is the oscillatoricity of the solutions of the differential equation (c_0) in (a, b).

REMARK 2.25. If w, b are given functions such that

$$\int_{x_0}^{x} [b(t)w^2(t) - w'''(t)w(t)]\, dt$$

is a bounded function of $x \in (a, b)$, then oscillatoricity of the solutions of (c_0) depends on an appropriate choice of the parameter $A_0(x_0)$.

LEMMA 2.18. Let $A_0(x) \leqq 0$ for $x \in (a, b)$. No solution u of the differential equation

$$u''' + 2A_0(x)u' = 0 \tag{2.80}$$

with $u(\alpha) = u'(\alpha) = 0$, $u''(\alpha) \neq 0$, $a < \alpha < b$, can have another zero, nor a zero of its first derivative in (a, b).

PROOF. Since $A_0(x) \leqq 0$, every solution of the differential equation

$$v'' + 2A_0(x)v = 0$$

has evidently at most one zero in (a, b). Let $v(x)$ now have the property $v(\alpha) = 0$, $v'(\alpha) \neq 0$. Then $u(x) = \int_{\alpha}^{x} v(t)\, dt$ is a solution of the differential equation (2.80) with the properties claimed for it, so the lemma is proved. □

REMARK 2.26. Lemma 2.3 gives a relation between solutions of the differential equations (a) and (a_1). If we take the differential equation (2.80) instead of (a_1) in this relation, then by Lemma 2.18 the function $W(x, t)$ is a solution of (2.80) (with t fixed) satisfying

$$W(x, z) \geqq 0 \text{ and } W'_x(x, t) \geqq 0 \text{ for } x \geqq t \in (a, b).$$

LEMMA 2.19. Let $w(x) \neq 0$ and its third derivative be continuous functions of $x \in (a, b)$. Let $b(x)$ be a continuous function oscillatory in (a, b) (or a function having finitely many simple zeros in (a, b), respectively) and let

$$\int_{x_0}^{x} [b(t)w^2(t) - w'''(t)w(t)]\, dt, \quad a < x_0 < b,$$

be a bounded function of $x \in (a, b)$. Let A_0, A_0' be given by (2.78), (2.79) and assume that A_0, A_0' and $A_0' - b$ are non-positive functions of $x \in (a, b)$ and that every solution of the differential equation (c_0) has at most one zero in (a, b). Then every solution of the differential equation (a_0) has at most two zeros or one double zero in (a, b).

PROOF. Lemma 2.19 will be proved if we show that no solution of (b_0) having a double zero at $\alpha \in (a, b)$ can have another zero to the right of α. Then the band at α contains at least one solution of (c_0) which has no other zero for $x > \alpha \in (a, b)$; therefore no solution belonging to the band at α has more than one zero to the right of α (except the solution having a double zero at α).

Let, therefore, z be a solution of the differential equation (b_0) with $z(\alpha) = z'(\alpha) = 0$, $z''(\alpha) > 0$ and let u be a solution of (2.80) satisfying $u(\alpha) = u'(\alpha) = 0$, $u''(\alpha) = z(\alpha) > 0$.

Lemma 2.3 implies that

$$z(x) = u(x) - \int_\alpha^x [A_0'(t) - b(t)]\, W(x, t) z(t)\, dt\,. \tag{2.81}$$

From the hypotheses of Lemma 2.19 and from Lemma 2.18 it follows that $u(x) > 0$ for $x > \alpha$, $A_0'(t) - b(t) \leqq 0$, $W(x, t) \geqq 0$ for $x \geqq t \geqq \alpha \in (a, b)$. Hence $0 < u(x) \leqq z(x)$ for $x > \alpha$. □

REMARK 2.27. Lemma 2.18 gives a fairly general construction of third order differential equations if b is an oscillatory function in (a, b) and the equation (a_0) is disconjugate in (a, b).

COROLLARY 2.17. Let $w = e^{-x}$. Then

$$A_0(x) = e^{2x} \left[A_0(x_0) + \int_{x_0}^x [b(t) + 1]\, e^{-2t}\, dt \right],$$

$$A_0'(x) = 2A_0(x) + b(x) + 1$$

and the differential equation (c_0) takes the form

$$y'' + y' + [1 + 2A_0(x)]\, y = 0\,.$$

The hypotheses of Lemma 2.18 are satisfied, for example, if the function $b(x)$ is bounded and the constant $A_0(x_0)$ is negative and sufficiently small.

LEMMA 2.20. Let the hypotheses of Lemma 2.18 be satisfied and, moreover, let $A_0'(x)+b(x)\geqq 0$ for $x\in(a, b)$. Then the derivative of a solution of (a_0) having a double zero at $\alpha\in(a, b)$ has no zero to the right of α.

PROOF. Apply Lemma 2.3 to the equations (2.80) and (a_0). Let y_1 be a solution of (a_0) with a double zero at α and let $y_1''(\alpha)>0$. Let u be a solution of (2.80) with $u(\alpha)=u'(\alpha)=0$, $u''(\alpha)=y_1''(\alpha)$. Lemma 2.3 gives

$$y_1(x)=u(x)-\int_\alpha^x [A_0'(t)+b(t)]\, W(x, t)\, y_1(t)\, dt\,.$$

After differentiating the last equality we have

$$y_1'(x)=u'(x)-\int_\alpha^x [A_0'(t)+b(t)]\, W_x'(x, t) y_1(t)\, dt,$$

whence the assertion follows, because $u'(x)\geqq 0$ for $x\geqq\alpha$, $W_x'(x, t)\geqq 0$ for $\alpha\leqq t\leqq x\in(a, b)$ by Remark 2.26.

REMARK 2.28. From Lemma 2.18, after differentiating (2.81), we get a relation implying an analogous assertion for the derivative of a solution of the differential equation (b_0) having a double zero at α.

THEOREM 2.66. Let the hypotheses of Lemma 2.19 be satisfied and, moreover, let $A(x)\leqq A_0(x)$, $A'(x)\leqq A_0'(x)$ for $x\in(a, b)$. Then every solution of the differential equation (a) has two zeros or one double zero, i.e. the differential equation (a) is disconjugate in (a, b).

PROOF. It is sufficient to show that no solution y_1 of the differential equation (a) having a double zero at $\alpha\in(a, b)$ has another zero and that no solution z_1 of the differential equation (b) having a double zero at α has another zero. We prove the statement concerning $y_1(x)$; for $z_1(x)$, the proof is analogous.

Apply Lemma 2.3 to the equations (a_0) and (a); this gives

$$y_1(x)=y_1^0(x)-2\int_\alpha^x [A(t)-A_0(t)]\, W(x,t)y_1'(t)\,dt-$$

$$-\int_\alpha^x [A'(t)-A_0'(t)]\, W(x,t)y_1(t)\,dt\,,$$

where y_1^0 is a solution of (a_0) satisfying at α the conditions $y_1^0(\alpha)=y_1^{0\prime}(\alpha)=0$, $y_1^{0\prime\prime}(\alpha)=y_1''(\alpha)$, $W(x,t)$ for $t\leqq x$ is a solution of (a_0) having a double zero at t, and $W(x,t)\geqq 0$, $W_x'(x,t)\geqq 0$ for $t\leqq x$ by Lemma 2.20.

Differentiating $y_1(x)$, we obtain

$$y_1'(x)=y_1^{0\prime}(x)-2\int_\alpha^x [A(t)-A_0(t)]\, W_x'(x,t)y_1'(t)\,dt-$$

$$-\int_\alpha^x [A'(t)-A_0'(t)]\, W_x'(x,t)y_1(t)\,dt\,.$$

It follows that $y_1'(x)$ has no zero to the right of α. The properties of bands (the differential equation (a) being in class I) imply the assertion of Theorem 2.66 □

COROLLARY 2.18. Using the formulae for $A_0(x)$ and $A_0'(x)$ in Corollary 2.17, we can easily verify that corresponding to every bounded oscillatory function b there exists a constant $-k^2$, $k>0$, such that the differential equation (a) is disconjugate in (a, b) whenever $A(x)\leqq -k^2\,e^{2x}$, $A'(x)\leqq -k^2\,e^{2x}$ for $x\in(a, b)$.

Another result fitting in this section is the criterion for the differential equation (a) to be disconjugate established by Ráb [117]. The proof is based on the so-called Mammana theorem (Mammana [94]) concerning decomposition of a third order operator into three operators of the first order. Mammana's theorem states the following:

A necessary and sufficient condition for the left-hand side of the differential equation (a) to decompose, in $\langle\alpha,\beta\rangle$, into a symbolic product of three linear differential operators (i.e. in order that every solution of (a) have at most two zeros or one double zero in $\langle\alpha,\beta\rangle$) is that the solutions of (a) and (b) having a double zero at α have no other zero in $\langle\alpha,\beta\rangle$.

We are not going to discuss this question in more detail. Instead in order to prove Ráb's theorem, we shall use the following lemma and the properties of bands.

LEMMA 2.21. Let y_1 be a solution of the differential equation (a) with $y_1(\alpha)=y_1'(\alpha)=0$, $y_1''(\alpha)>0$, and let z_1 be a solution of the differential equation (b) with $z_1(\alpha)=z_1'(\alpha)=0$, $z_1''(\alpha)>0$, $a<\alpha<b$. Also, let $y_1(x)>0$, $z_1(x)>0$ for $x>\alpha$. Then the differential equations (a) and (b) are in class II in (α, b).

PROOF. We shall prove that the differential equation (a) belongs to class II. For the equation (b), the proof is analogous.

Let y_1 and z_1 have the assumed properties, let $\bar{y}_1$ be a solution of (a) with $\bar{y}_1(\beta)=\bar{y}_1'(\beta)=0$, $\bar{y}_1''(\beta)>0$ and let $\bar{y}_1(x_1)=0$, $a<\alpha<\beta<x_1<b$, where x_1 is the first zero of $\bar{y}_1$ to the right of β. By Lemma 1.2, there exist a constant $c>0$ and a point ξ, $\beta<\xi<x_1$, such that the function $y=\bar{y}_1-cy_1$ has a double zero at ξ. This, however, is possible only if $z(\xi)=\bar{y}_1(\xi)y_1'(\xi)-\bar{y}_1'(\xi)y_1(\xi)=0$, where $z(x)=\bar{y}_1(x)y_1'(x)-\bar{y}_1'(x)y_1(x)$ is a solution of the differential equation (b) with $z(\alpha)=z(\beta)=z(\xi)=0$. Therefore, the function z belongs to the band of the differential equation (b) at α together with the solution $z_1(x)\neq 0$ for $x>\alpha$. Thus, z can have at most one zero to the right of α, due to the regularity of the band since $y_1(x)\neq 0$ for $x>\alpha$. □

THEOREM 2.67. For $a<\alpha<x<b$, let

$$A(x)+\sup_{\xi\in(\alpha,x)}\left|\int_\xi^x b(t)\,\mathrm{d}t\right|\leq\frac{1}{2(x-\alpha)^2}. \tag{2.82}$$

Then every solution of the differential equation (a) has at most two zeros or one double zero in $\langle\alpha, b)$.

PROOF. In view of Lemma 2.21, it is sufficient to show that no solution y_1 of (a) and no solution z_1 of (b) satisfying $y_1(\alpha)=y_1'(\alpha)=0$, $y_1''(\alpha)>0$, $z_1(\alpha)=z_1'(\alpha)=0$, $z_1''(\alpha)>0$ can have another zero in (α, b).

Again, it is enough to prove the assertion for y_1, since the proof for z_1 is analogous. The integral identity (1.6) for the solution y_1 takes the form

$$y_1y_1''-\frac{1}{2}y_1'^2+Ay_1^2=-\int_\alpha^x b(t)y_1^2(t)\,dt\,. \tag{2.83}$$

Since $y_1''(\alpha)>0$, we have $y_1''(x)>0$ in some right neighbourhood of α and $y_1(x)$ is a convex function in that neighbourhood. We shall show that $y_1(x)$ is convex in the whole interval $\langle\alpha, b)$. Suppose the contrary, and denote by x_1 the infimum of all x for which $y_1''(x)\leqq 0$. The number x_1 lies in the interior of $\langle\alpha, b)$ and $y_1''(x_1)=0$. Putting $x=x_1$ in (2.83), we obtain

$$\frac{1}{2}y_1'^2(x_1)=A(x_1)y_1^2(x_1)+\int_\alpha^{x_1} b(t)y_1^2(t)\,dt\,. \tag{2.84}$$

By the mean value theorem, there exists some $\xi_1\in(\alpha, x_1)$ such that $y_1(x_1)-y_1(\alpha)=(x_1-\alpha)y_1'(\xi_1)$, i.e.

$$y_1'(\xi_1)=\frac{y_1(x_1)}{x_1-\alpha}\,. \tag{2.85}$$

Since $y_1''(x)>0$ for $x\in(\alpha, x_1)$, the function $y_1'(x)$ is increasing in that interval and hence, by (2.85),

$$y_1'(x_1)>\frac{y(x_1)}{x_1-\alpha}\,.$$

Applying this inequality to (2.84) yields

$$\frac{1}{2}\frac{y_1^2(x_1)}{(x_1-\alpha)^2}<\frac{1}{2}y'^2(x_1)=A(x_1)y_1^2(x_1)+\int_\alpha^{x_1} b(t)y_1^2(t)\,dt,$$

hence

$$\frac{1}{2(x_1-\alpha)^2}<A(x_1)+\int_\alpha^{x_1} b(t)\frac{y_1^2(t)}{y_1^2(x_1)}\,dt\,. \tag{2.86}$$

The function

$$\Psi(t)=\frac{y_1(t)}{y_1(x_1)}$$

is increasing in (α, x_1) and $\Psi(\alpha)=0$, $\Psi(x_1)=1$. Therefore, by the second mean value theorem of the integral calculus, there is some $\xi\in(\alpha, x_1)$ such that

$$\int_\alpha^{x_1} b(t)\frac{y_1^2(t)}{y_1^2(x_1)}\,dt=\int_\xi^{x_1} b(t)\,dt\,.$$

Then (2.86) reduces to

$$\frac{1}{2(x_1-\alpha)^2}<A(x_1)+\int_\xi^{x_1} b(t)\,dt\,.$$

It follows that

$$\frac{1}{2(x_1-\alpha)^2}<A(x_1)+\sup_{\xi\in(\alpha,\,x_1)}\int_\xi^{x_1} b(t)\,dt\,,$$

that is,

$$\frac{1}{2(x_1-\alpha)^2}<A(x_1)+\sup_{\xi\in(\alpha,\,x_1)}\left|\int_\xi^{x_1} b(t)\,dt\right|,$$

which contradicts the assumption. Therefore, $y_1(x)$ is a convex function in $\langle\alpha, b)$, and the theorem is proved. □

13. The Differential Equation (a) Having All Solutions Oscillatory in a Given Interval

Mammana in [94] conjectured that a linear differential equation of the third order must always have a solution without zeros in the interval under consideration. However, Ascoli [3] has constructed an example of a third order equation with all solutions having at least one zero in the interval. Sansone [125] has constructed a third order equation of the form

$$y'''+Q(x)y'+Q'(x)y=0 \tag{2.87}$$

whose solutions all have a prescribed number of zeros in the given interval. In the paper by Greguš [39], sufficient conditions on Q are formulated to ensure that every solution of (2.87) has infinitely many zeros in $(-\infty, \infty)$.

The construction in Sansone's paper [125] is as follows.

Let $y=y(x)$ be a continuous function with continuous finite derivatives up to (a, b) and including the fourth order let

$$a<\alpha=\alpha_0<\alpha_1<\ldots<\alpha_{4N+2}<\alpha_{4N+3}=\beta<b$$

be its zeros and let

$$y(\alpha_k)=0,\ \ y'(\alpha_k)\neq 0,\ \ y''(\alpha_k)=0,$$
$$k=0,1,\dots,4N+3\,, \tag{2.88}$$

where $N\geqq 0$ is an integer.

Besides, let $(-1)^k y(x)>0$ for $\alpha_k<x<\alpha_{k+1}$, $k=0,1,\dots,4N+2$, and hence

$$(-1)^k y'(\alpha_k)>0\,,\quad k=0,1,\dots,4N+3\,. \tag{2.89}$$

Put

$$(-1)^k\int_{\alpha_k}^{\alpha_{k+1}} y(t)\,\mathrm{d}t=I_{k+1} \tag{2.90}$$

and assume, moreover, that

$$I_{2k-1}+I_{2k+1}<I_{2k}\,,\quad k=1,2,\dots,2N+1\,. \tag{2.91}$$

The function $y''(x)/y(x)$ has finite limits at the points α_k since

$$\lim_{x\to\alpha_k}\frac{y''(x)}{y(x)}=\frac{y'''(\alpha_k)}{y'(\alpha_k)}\,.$$

For $x\in\langle\alpha,\beta\rangle$, put

$$Q(x)=-\frac{y''(x)}{y(x)}\ \text{ for }\ x\neq\alpha_k\quad\text{and}$$
$$Q(x)=\frac{y'''(\alpha_k)}{y'(\alpha_k)}\ \text{ for }\ x=\alpha_k\,. \tag{2.92}$$

The function $Q'(x)$ evidently exists and is continuous for $x\neq\alpha_k$. We now show that $Q'(x)$ is continuous also at the points $x=\alpha_k$. We have

$$\frac{Q(x)-Q(\alpha_k)}{x-\alpha_k}=\frac{-\dfrac{y''(x)}{y(x)}+\dfrac{y'''(\alpha_k)}{y'(\alpha_k)}}{x-\alpha_k}\quad\lim_{x\to\alpha_k}\frac{Q(x)-Q(\alpha_k)}{x-\alpha_k}=$$
$$=\lim_{x\to\alpha_k}\frac{-y'''(x)y(x)+y''(x)y'(x)}{y^2(x)}=$$
$$=-\frac{1}{2}\lim_{x\to\alpha_k}\frac{y^{(4)}(x)y(x)-y''^2(x)}{y(x)y'(x)}=$$

$$= -\frac{1}{2}\frac{y^{(4)}(\alpha_k)}{y'(\alpha_k)} + \frac{1}{y'(\alpha_k)}\lim_{x\to\alpha_k}\frac{y''(x)}{y(x)}\,y''(x) =$$

$$= -\frac{1}{2}\frac{y^{(4)}(\alpha_k)}{y'(\alpha_k)}.$$

Since for $x \neq \alpha_k$ we have

$$Q'(x) = -\frac{y'''(x)y(x) - y''(x)y'(x)}{y^2(x)},$$

the function $Q'(x)$ is continuous in $\langle \alpha, \beta \rangle$.

The differential equation (2.87) is obtained by termwise differentiation of the second order differential equation

$$y'' + Q(x)y = 0 .$$

Evidently, y is a solution of this when Q is given by (2.92). Let z be a solution of the same equation, independent of y, and let

$$zy' - z'y = 1 . \tag{2.93}$$

The general solution of the second order differential equation

$$Y'' + Q(x)Y = 1$$

is

$$Y = y(x)\int_\alpha^x z\,dt - z(x)\int_\alpha^x y\,dt + c_1 y(x) + c_2 z(x),$$

where c_1, c_2 are arbitrary constants (this solution being obtained by the method of variation of constants). We proceed to show that Y has at least $N+1$ zeros in (α, β). This will prove that the general solution of (2.87) has at least $N+1$ zeros in (α, β), because cY is the general solution of (2.87).

Clearly,

$$Y(\alpha_k) = -z(\alpha_k)\int_\alpha^{\alpha_k} y\,dt + c_2 z(\alpha_k) =$$

$$= -z(\alpha_k)\left[\int_\alpha^{\alpha_k} y\,dt - c_2\right].$$

From (2.93) it follows that $z(\alpha_k)y'(\alpha_k)=1$ and hence, in view of (2.89), the sign of $z(\alpha_k)$ is $(-1)^{k+1}$, and since α_k ranges over α_0, $\alpha_1, \dots, \alpha_{4N+3}$, $z(x)$ successively changes its sign at these points.

Assume that the four members

$$Y(\alpha_{2k})\,,\quad Y(\alpha_{2k+1})\,,\quad Y(\alpha_{2k+2})\,,\quad Y(\alpha_{2k+3})$$

have the same sign. We can see that this is not possible, for that would imply that the expressions

$$\int_\alpha^{\alpha_{2k-1}} y\,dt - c_2 + I_{2k}, \quad \int_\alpha^{\alpha_{2k-1}} y\,dt - c_2 + I_{2k} - I_{2k+1},$$

$$\int_\alpha^{\alpha_{2k-1}} y\,dt - c_2 + I_{2k} - I_{2k+1} - I_{2k+2},$$

$$\int_\alpha^{\alpha_{2k-1}} y\,dt - c_2 + I_{2k+1} + I_{2k+2} - I_{2k+3}$$

change their signs, hence that

$$\int_\alpha^{\alpha_{2k-1}} y\,dt - c_2 - I_{2k} > 0\,,$$

$$\int_\alpha^{\alpha_{2k-1}} y\,dt - c_2 + I_{2k} + I_{2k+1} < 0\,,$$

$$\int_\alpha^{\alpha_{2k-1}} y\,dt - c_2 + I_{2k} - I_{2k+1} + I_{2k+2} > 0\,,$$

$$\int_\alpha^{\alpha_{2k-1}} y\,dt - c_2 + I_{2k} - I_{2k+1} + I_{2k+2} - I_{2k+3} < 0\,,$$

and finally that

$$0 < \int_\alpha^{\alpha_{2k-1}} y\,dt - c_2 + I_{2k} < I_{2k+1} - I_{2k+2} + I_{2k+3}\,,$$

which contradicts the assumption (2.91). Therefore, Y must have at least one zero between every two numbers α_0 and α_3, α_4 and α_7, ... α_{4N} and α_{4N+3}. Thus, Y has at least $N+1$ zeros in (α, β), and the construction is complete. □

REMARK 2.29. If y has infinitely many zeros in $\langle a, b)$ and all the remaining assumptions about y hold, then it follows from the construction that every solution of the differential equation (2.87) so constructed has infinitely many zeros in $\langle a, b)$, i.e. it oscillates in (a, b).

Consider again the second order differential equation

$$y'' + Q(x)y = 0\,, \tag{2.94}$$

where $Q(x)$ and $Q'(x)$ are continuous functions in $(-\infty, \infty)$. Differentiation of the equation (2.94) gives the differential equation (2.87); it is an equation of the type (a) with $b(x) = 1/2Q'(x)$. The adjoint equation to (2.87) is

$$z''' + Q(x)z' = 0\,. \tag{2.95}$$

LEMMA 2.22. Let $Q(x) > 0$ for $x \in (-\infty, \infty)$ and let $Q'(x) > 0$ $[Q'(x) < 0]$ for $x \in (-\infty, \infty)$.

Then the solutions of (2.87) have the following properties:

a) All the solutions of (2.87) belonging to the band at α, $-\infty < \alpha < \infty$, i.e. with $y(\alpha) = 0$, oscillate for $x > \alpha [x < \alpha]$ and their zeros separate each other in (α, ∞) $[(-\infty, \alpha)]$.

b) The band at α contains one solution (up to linear dependence) which is, at the same time, a solution of (2.94). Moreover, the band at α includes the set of solutions of (2.87) each of which has a double zero to the left (right) of α. The double zero of each of these solutions is precisely at a zero of the adjoint differential equation (2.95) having a double zero at α. (The set may be empty if the solution of (2.95) under consideration has no zeros to the left [right] of α.) Exactly one solution (up to linear dependence) from every band at any point on the left [right] of α belongs also to the band at α. Finally, the band at α contains infinitely many solutions of (2.87) which have no zero to the left [right] of α, while only one of them (again, up to linear dependence) has a double zero at α.

c) Every solution of (2.87) belongs to some band or has no zero in $(-\infty, \infty)$. The solutions without zeros are in the set of solutions $y = c_1 y_1 + c_3(-t y_2 + y_3)$, where y_1, y_2, y_3 is a fundamental system of solutions of (2.87) and t is a parameter ranging over some finite interval $\langle t_1, t_2 \rangle$.

REMARK 2.30. The notion of oscillatoricity of a solution of (2.87) or (a) in a given interval (a, b) was defined in such a manner that the point b is a limit point of the zeros of the solutions. By oscillatoricity of a solution to the left of α, $a<\alpha<b$, we mean that the solution has infinitely many zeros in $(a, \alpha\rangle$ and a is their limit point. In our case, $a=-\infty$. Further concepts may be introduced analogously, such as oscillatoricity of a differential equation to the left of α and non-oscillatoricity.

PROOF OF LEMMA 2.22. We shall prove the lemma in the case $Q'(x)>0$. For $Q'(x)<0$, the proof is analogous.

a) Let $-\infty<\alpha<\infty$. Let y_1, y_2 be a fundamental system of solutions of the differential equation (2.94) with wronskian equal to one. Then, evidently,

$$y_3(x)=-y_1(x)\int_\alpha^x y_2(t)\,dt+y_2(x)\int_\alpha^x y_1(t)\,dt$$

is a solution of (2.87) with $y_3(\alpha)=y_3'(\alpha)=0$, $y_3''(\alpha)\neq 0$ and y_1, y_2, y_3 form a fundamental set of solutions of (2.87) (see Corollary 1.4 and Remark 1.5). Now construct the solution $\bar{y}_1(x)=-y_1(\alpha)y_2(x)+y_2(\alpha)y_1(x)$; obviously, $\bar{y}_1(\alpha)=\bar{y}_1''(\alpha)=0$, $\bar{y}_1'(\alpha)\neq 0$. The set of solutions $y(x)=c_1\bar{y}_1(x)+c_2y_3(x)$, where c_1 and c_2 are arbitrary constants, has the property $y(\alpha)=0$. Consider the wronskian $w(\bar{y}_1, y_3)=w(x)$. Evidently,

$$w(x)=y_2(\alpha)\int_\alpha^x y_1(t)\,dt-y_1(\alpha)\int_\alpha^x y_2(t)\,dt\,.$$

The function $w(x)$ satisfies the adjoint differential equation (2.95) and also the conditions $w(\alpha)=w'(\alpha)=0$. From the integral identity (1.8) for the solution w (with $2A=Q$, $b=1/2Q'$) it follows that $w(x)\neq 0$ for $x>\alpha$.

The set of solutions $y=c_1\bar{y}_1+c_2y_3$ satisfies, for $x>\alpha$, the differential equation (c) (with $2A=Q$), whose solutions oscillate for $x>\alpha$, because the solution $\bar{y}_1$ does so. Since $w(x)\neq 0$ for $x>\alpha$, the zeros of the set $y=c_1\bar{y}_1+c_2y_3$ separate each other for $x>\alpha$.

b) The band at α evidently contains one solution (up to linear dependence) of the equation (2.94), namely the solution $\bar{y}_1$. Now we prove the second part of the assertion b). In order that $y(x)=c_1\bar{y}_1(x)+c_2y_3(x)$ should have a double zero at some $x<\alpha$, the constants c_1, c_2 must be non-zero and such that

$$c_1\bar{y}_1(x)+c_2y_3(x)=0\,,$$
$$c_1\bar{y}_1'(x)+c_2y_3'(x)=0\,.$$

The last two equations have a non-zero solution c_1, c_2 only at those points x at which

$$\begin{vmatrix}\bar{y}_1, & y_3\\ \bar{y}_1', & y_3'\end{vmatrix}=y_2(\alpha)\int_\alpha^x y_1(t)\,\mathrm{d}t-y_2(\alpha)\int_\alpha^x y_2(t)\,\mathrm{d}t=0\,.$$

The solution

$$w(x)=y_2(\alpha)\int_\alpha^x y_1(t)\,\mathrm{d}t-y_1(\alpha)\int_\alpha^x y_2(t)\,\mathrm{d}t$$

satisfies the differential equation (2.95) and has a double zero at α.

In the set of solutions $y=c_1\bar{y}_1+c_2y_3$, it is always possible to choose c_1 and c_2 so that $y(x)=0$ at any point $x<\alpha$. Thus, from each band on the left of α there arises a solution which, at the same time, belongs to the band at α. Uniqueness (up to linear dependence) is implied by Theorem 1.2.

We still have to prove that the band at α contains infinitely many solutions having no zero to the left of α.

It follows from the integral identity (1.6) that $y_3(x)\neq 0$ for $x<\alpha$. We now show that the constants c_1, c_2 can be chosen so that $c_1\bar{y}_1+c_2y_2>0$ for $x<\alpha$. Let (α_1,α), $\alpha_1<\alpha$, be an interval in which every solution of the differential equation (2.94) has at least two zeros. (If the solutions of (2.94) do not oscillate for $x<\alpha$, the assertion is obvious.) It suffices to prove that c_1 and c_2 can be chosen so that $c_1\bar{y}_1+c_2y_3>0$ for $x\in(\alpha_1,\alpha)$. In fact, if we had $c_1\bar{y}_1+c_2y_3=0$ for $x_1\leqq\alpha$, then $c_1\bar{y}_1+c_2y_3$ would belong to the band at x_1, which contains also a solution of (2.94) having at least two zeros in (α_1,α), and hence $c_1\bar{y}_1+c_2y_3$ would necessarily have at least one zero in (α_1,α).

Without loss of generality we may assume that $y_3(x)\geqq 0$ for $x<\alpha$. Let $\bar{x}$ be the first zero of $\bar{y}_1$ to the left of α. Two cases may arise, either that $\bar{y}_1(x)>0$ in $(\bar{x}, \alpha)$, or $\bar{y}_1(x)<0$ in $(\bar{x}, \alpha)$. Let $\bar{y}_1(x)>0$ in $(\bar{x}, \alpha)$. If both c_1 and c_2 are positive, then $c_1\bar{y}_1+c_2y_3>0$ for $x\in(\bar{x}, \alpha)$. However, they must be chosen so that $c_1\bar{y}_1+c_2y_3>0$ for $x\in(\alpha_1, \alpha)$, $\alpha_1<\bar{x}$. Let $\beta>0$ be the infimum of y_3 over $\langle\alpha_1, \bar{x})$ and let $|\bar{y}_1|<\gamma$, $\gamma>0$ for $x\in(\alpha_1, \alpha)$. Then it suffices to choose c_1, c_2 with $c_2\beta>c_1\gamma$. This can always be done, indeed in infinitely many ways. If $\bar{y}_1<0$ for $x\in(\bar{x}, \alpha_1)$, then c_1 can be chosen negative and c_2 positive, such that $c_2\beta>-c_1\gamma$.

Similarly we could find c_1, c_2 with $c_1\bar{y}_1+c_2y_3<0$ for $x\in(\alpha_1, \alpha)$. Thus, the assertion b) is proved.

c) Further, choose y_1 and y_2 with $y_1(\alpha)=0$, $y_2(\alpha)>0$, $y_1'(\alpha)<0$. The general solution of the differential equation (2.87) is of the form

$$y=c_1y_1+c_2y_2+c_3y_3\ .$$

We shall prove the assertion c) in two parts.

1. Let $\operatorname{sgn} c_2\neq\operatorname{sgn} c_3$.

We show that every solution with the above property belongs to some band. Suppose the contrary, i.e. that $y(x)\neq0$ for $x>\alpha$. Consider

$$\left(\frac{y_1(x)}{y(x)}\right)'=-\frac{c_2+c_3\displaystyle\int_\alpha^x y_1(t)\,dt}{y^2}\ . \tag{2.96}$$

The function $z(x)\overset{\mathrm{d}}{=}c_2+c_3\displaystyle\int_\alpha^x y_1(t)\,dt$ is evidently a solution of the differential equation (2.95) satisfying $z(\alpha)=c_2$, $z'(\alpha)=0$, $z''(\alpha)=c_3y_1'(\alpha)$. The integral identity (1.8) for the solution z reads

$$zz''-\frac{1}{2}z'^2+\frac{1}{2}Qz^2-\frac{1}{2}\int_\alpha^x Q'z^2\,dt=$$

$$=c_2c_3y_1'(\alpha)+\frac{1}{2}Q(\alpha)x_2^2.$$

Suppose $z(x_1)=0$ for $x_1>\alpha$. At x_1, the integral identity implies that

$$-\frac{1}{2}z'^2(x_1)-\frac{1}{2}\int_\alpha^x Q'z^2\,dt=c_2c_3y_1'(\alpha)+\frac{1}{2}Q(\alpha)c_2^2.$$

But this is a contradiction, therefore $z(x)$ has no zero to the right of α.

Let $\alpha<x_1<x_2$ be zeros of y. Integrating (2.96) from x_1 to x_2, we

obtain zero on the left-hand side and an expression not equal to zero on the right-hand side. This contradiction proves that y has at least one zero to the right of α.

2. Let $\operatorname{sgn} c_2 = \operatorname{sgn} c_3$.

Let $\alpha < x_1 < x_2 < x_3 < \ldots < x_k \ldots$ be all the zeros of y_1 to the right of α. $\int_\alpha^x y_1\, dt$ is a solution of (2.95) having a double zero at α. The integral identity (1.8) for this solution clearly implies that $\int_\alpha^x y_1\, dt \neq 0$ for $x > \alpha$. Moreover, $\int_\alpha^x y_1\, dt < 0$ for $x > \alpha$, because $\int_\alpha^{x_1} y_1\, dt < 0$ while the first derivative of $\int_\alpha^x y_1\, dt$ oscillates in $\langle \alpha, \infty)$. This argument implies that

$$\left| \int_\alpha^{x_1} y_1\, dt \right| > \left| \int_{x_1}^{x_2} y_1\, dt \right|$$

and also

$$\left| \int_{x_{k-1}}^{x_k} y_1\, dt \right| > \left| \int_{x_k}^{x_{k+1}} y_1\, dt \right|$$

for $k = 2, 3, \ldots$.

Evidently, $\int_\alpha^x y_1\, dt$ has a minimum at each x_{2k-1} and a maximum at each x_{2k}, while the minima of $\int_\alpha^x y_1\, dt$ increase, and the maxima decrease, as k increases. Since no minimum can be greater than any maximum, the minima converge to some number t_1 and the maxima converge to some t_2, $t_1 \leqq t_2$.

Since $\operatorname{sgn} c_2 = \operatorname{sgn} c_3$ and $\int_\alpha^x y_1\, dt < 0$ for $y > \alpha$, we have $\operatorname{sgn} c_2 \neq \operatorname{sgn} c_3 \int_\alpha^x y_1\, dt$ for $x > \alpha$.

Obviously, $\lim_{k\to\infty} c_3 \int_\alpha^{x_{2k-1}} y_1\, dt = c_3 t_1$ and $\lim_{k\to\infty} c_3 \int_\alpha^{x_{2k}} y_1\, dt = c_3 t_2$. If $c_2 < -c_3 t_1$ or $c_2 > -c_3 t_2$, then for some sufficiently great k we have

$c_2+c_3\int_\alpha^x y_1\,dt\neq 0$ for $x\in(x_k, x_{k+1})$. Integrating (2.96) from x_k to x_{k+1} again yields a contradiction, and so y must have at least one zero for $x>\alpha$.

The remaining case is $c_2=-c_3t$, where $t_1\leqq t\leqq t_2<0$. In this case we get the set of solutions $y=c_1y_1+c_3(-ty_2+y_3)$, among which there are solutions having no zeros at all on the real axis, because all the other solutions oscillate for $x>\alpha$. According to Theorem 1.13, the differential equation (2.87) with the Laguerre invariant $b(x)=1/2Q'(x)>0$ for $x\in(-\infty,\infty)$ has at least one solution without zeros in $(-\infty, \infty)$. Thus, the lemma is proved. □

THEOREM 2.68. Let $Q(x)>0$ for $x\in(-\infty, \infty)$, $Q'(x)<0$ for $x<\alpha$, $Q'(\alpha)=0$, $Q'(x)>0$ for $x>\alpha$, $-\infty<\alpha<\infty$. Moreover, let every solution of the differential equation (2.94) have the property that $y\to 0$ as $x\to\pm\infty$ and assume that $Q(x)$ is not symmetrical with respect to the axis perpendicular to the y-axis through α. Then every solution of the differential equation (2.87) has infinitely many zeros in $(-\infty, \infty)$.

PROOF. Let y_1, y_2 be a fundamental set of solutions of (2.94) such that their wronskian equals one and $y_1(\alpha)=0$, $y_1'(\alpha)<0$, $y_2(\alpha)>0$. Let $y_3=-y_1\int_\alpha^x y_2\,dt+y_2\int_\alpha^x y_1\,dt$. From the above reasoning we know that y_1, y_2, y_3 is a fundamental system of solutions of (2.87). The properties of Q imply that the integrals of (2.44) oscillate both to the left and to the right of α.

The general solution of (2.87) is $y=c_1y_1+c_2y_2+c_3y_3$. By similar reasoning to that used for Lemma 2.21, assertion c) we can prove that in the case sgn $c_2\neq$ sgn c_3 the solutions of (2.87) with the specified property oscillate to the left as well as to the right of α.

Now let sgn $c_2=$ sgn c_3.

The integral identity (1.8) implies that neither to the left nor to the right of α has $\int_\alpha^x y_1\,dt$ any zeros.

Let $\ldots x_{-3}<x_{-2}<x_{-1}<\alpha<x_1<x_2<x_3<\ldots$ be all the zeros of y_1. By similar reasoning to that used in the proof of Lemma 2.21, we deduce that $\int_\alpha^x y_1\,dt$ attains its minima at x_{2k-1} or $x_{-(2k-1)}$, and mixima at x_{2k} or x_{-2k}. With increasing k, the maxima of $\int_\alpha^x y_1\,dt$ to the right (left) of α decrease and the minima increase. Since no minimum can be greater that any maximum (to the right or left of α, respectively) and since $y_1(x)\to 0$ as $x\to\pm\infty$, the minima and the maxima converge to some T_1 on the right of α and to T_2 on the left of α. As $Q(x)$ is not symmetrical with respect to $x=\alpha$, we have $T_1\neq T_2$.

Evidently, sgn $c_2\neq$ sgn $c_3\int_\alpha^x y_1\,dt$ both to the left and to the right of α. If $c_2\neq -c_3T_1$ and, at the same time, $c_2\neq c_3T_2$, then y oscillates to the left as well as to the right, which can be verified as in Lemma 2.22. If $c_2=-c_3T_1$, then $c_2\neq -c_3T_2$, hence the solution y oscillates in $(-\infty,\alpha)$. If $c_2=-c_3T_2$, then $c_2\neq -c_3T_1$, and thus y oscillates in (α,∞). □

THEOREM 2.69. Let $-\infty<\alpha<\infty$. Let the coefficients A, A', b of the differential equation (a) be continuous in $(-\infty,\infty)$ and suppose that in $\langle\alpha,\infty)$ they satisfy the hypotheses of Theorem 2.15. Moreover, let $b(\alpha+x)=-b(\alpha-x)$ and $A(\alpha+x)=A(\alpha-x)$. Besides, let A, b be such that the differential equation (a) has at least one oscillatory solution in $\langle\alpha,\infty)$. Then every solution of (a) is oscillatory both to the left and to the right of α, except two solutions (up to linear dependence), one of which oscillates to the left of α and has no zeros for $x>\alpha$, and the other oscillates to the right of α and has no zeros for $x<\alpha$.

PROOF. In order to simplify the proof, let $\alpha=0$. From the assumption that A is an even function and b an odd function it follows that whenever $y(x)$ is a solution of (a), then so is $y(-x)$. By Theorem 2.15, there exists exactly one solution (up to linear dependence) $y_1(x)$ of (a) such that $y_1(x)>0$ for $x\geqq 0$, $y_1'(0)<0$. The function $y_2(x)=y_1(-x)$ also satisfies the differential equation (a) and the condition $y_2(0)>0$, $y_2'(0)>0$, hence it oscillates for $x>0$, implying that y_1 oscillates for $x<0$.

The solutions y_1, y_2 are linearly independent and such that y_1 oscillates only on the left of the origin and y_2 oscillates on the right of the origin only. Every other solution of (a) must oscillate on both sides of the origin, otherwise it would necessarily be dependent on y_1 or y_2. □

§3. ASYMPTOTIC PROPERTIES OF SOLUTIONS OF THE DIFFERENTIAL EQUATIONS (a) AND (b)

In this section, we shall investigate asymptotic properties of solutions of the differential equations (a) and (b) without zeros in (a, ∞), then asymptotic properties of oscillatory solutions of the differential equations (a) and (b), and finally, asymptotic properties of all the solutions of the differential equation (a). The results presented are due to Zlámal [146], Ráb [114—116], Švec [137], Lazer [86], Greguš [47, 48] and Jones [71]. Asymptotic properties of the binomial equation of the third order were studied by Villari [143] and Švec [135].

1. *Asymptotic Properties of Solutions without Zeros of the Differential Equations (a) and (b)*

THEOREM 3.1 Let $A(x) \geqq m > 0$, $A'(x) + b(x) \geqq m$, $b(x) - A'(x) \geqq 0$ for every $x \in (a, \infty)$. Then every solution of the differential equation (a) is oscillatory in (a, ∞) except one solution y (up to linear dependence) for which $\lim_{x \to \infty} y(x) = 0$, $\lim_{x \to \infty} y'(x) = 0$ and y is in L^2.

PROOF. The hypotheses imply that $b(x) \geqq m > 0$. Also, $A(x) \geqq m$, hence by Theorem 2.60, the differential equation (a) is oscillatory in (a, ∞), i.e. every solution having a zero oscillates in (a, ∞). Thus, non-oscillatory solutions are non-zero in the whole interval (a, ∞).

Let, therefore, y be a solution without zeros (at least one such solution exists by Theorem 1.13) and let $y(x) > 0$ for $x \in (a, \infty)$. We show that it is impossible to have $y'(x) \geqq 0$ starting from some x. If $y'(x) \geqq 0$ for $x \geqq x_0 > a$, then $y(x) \geqq y(x_0) > 0$ for $x \geqq x_0$ and (a) implies that $y'''(x) \leqq -y(x_0)m$, which is a contradiction. Therefore, either $y'(x) \leqq 0$ beginning at some x or $y'(x)$ changes its sign infinitely many times. In the former case, y must be in L^2, otherwise we would have $\int_{x_0}^{x} b(t)y^2(t)\,dt \to \infty$ as $x \to \infty$ and it would follow from (1.6) for y that on the right, beginning at some x, $2yy'' - y'^2 + 2Ay^2 < 0$, so that

$$y^2\left[2\left(\frac{y'}{y}\right)' + \frac{y'^2}{y^2} + 2A\right] < 0, \text{ that is, } \left(\frac{y'}{y}\right)' < 0$$

for large x. Thus if x_1 is such that $y'(x_1) < 0$, then for $x > x_1$ we have

$$y(x) < k\, e^{[y'(x_1)/y(x_1)](x - x_1)},$$

where $k > 0$ is a constant, therefore y is indeed in L^2, i.e., $\int_{x_1}^{\infty} y^2(t) < \infty$. Hence, using the assumption that $y'(x) \leqq 0$, we conclude that $\lim_{x\to\infty} y(x) = 0$. Evidently, also, $\lim_{x\to\infty} y'(x) = 0$ since the integral identity (1.7) for y implies that $y''(x)$ is bounded above for large positive x.

If $y'(x)$ changes its sign infinitely many times, we proceed as follows. First of all, observe that the integral identity (1.7) implies that

$$y'' + 2Ay = k - \int_{x_1}^{x} [b(t) - A'(t)]y(t)\,dt \leqq k\,.$$

Moreover, y attains a local minimum infinitely many times. Let $\{x_n\}_{n=1}^{\infty}$ be a sequence of numbers at which y attains a minimum, i.e. we have $y(x_n) > 0$, $y'(x_n) = 0$, $y''(x_n) > 0$ for all x_n, $n = 1, 2, \ldots$. It follows from the integral identity (1.6) that $2F(y(x_n)) = 2y(x_n)y''(x_n) + 2A(x_n)y^2(x_n) > 0$. On the other hand, $F(y(x))$ is a decreasing function of x, because $b(x) \geqq m > 0$. Therefore, $F(y(x)) > 0$ for all x.

Hence

$$F(y(x_1)) - \int_{x_1}^{x} b(t)y^2(t)\,dt > 0, \quad b(t) \geqq m > 0\,,$$

implying that $\int_{x_1}^{\infty} y^2(t)\,dt < \infty$.

Further,

$$0 > -2F(y(x)) = y'^2 + 2Ay^2 - 2(y'' + 2Ay)y >$$
$$> y'^2 + 2my^2 - 2ky\ ,$$

i.e. simultaneously

$$my^2 - ky < 0\ , \quad y'^2 - 2ky < 0\ .$$

The former inequality implies that $y < k/m$, the latter that $|y'| < 2k/\sqrt{m}$, so $[y^2(x)]'$ is a bounded function. From this result and the fact that $y \in L^2$, we conclude that $\lim_{x\infty} y(x) = 0$. Then $y'^2 - 2ky < 0$ implies that $\lim_{x\to\infty} y'(x) = 0$.

Now we show that there is only one solution y without zeros with the property $y \to 0$, $y' \to 0$ as $x \to \infty$. Suppose the contrary. Let z be a solution without zeros, independent of y. Let y_1, y_2 be solutions of the differential equation (a) with $y_1(\alpha) = y_1'(\alpha) = 0$, $y_1''(\alpha) > 0$, $y_2(\alpha) = y_2''(\alpha) = 0$, $y_2'(\alpha) > 0$. Then $z = c_1y_1 + c_2y_2 + c_3y$. Let $Y = c_1y_1 + c_2y_2$, the solution Y is oscillatory in (a, ∞). Let $x_1 < x_2 < \ldots$ be zeros of Y. The integral identity (1.6) for Y implies

$$Y'^2(x_i) = Y'^2(x_1) + 2\int_{x_1}^{x_i} bY^2\, dt\ .$$

We therefore cannot have $\lim_{x\to\infty} z'(x) = 0$, so the theorem is proved. □

THEOREM 3.2. Let $b(x)$ have the property (v) in (a, ∞) and, moreover, let $A(x) \leqq 0$, $A'(x) + b(x) \leqq 0$ for $x \in (a, \infty)$. Then at least one solution of the differential equation (a) tends, together with its derivative, to ∞ as $x \to \infty$.

PROOF. By Remark 2.2, every solution of (a) has at most two zeros or one double zero in (a, ∞). We shall now show that every solution y of (a) with $y(\alpha) \geqq 0$, $y'(\alpha) \geqq 0$, $y''(\alpha) > 0$, $a < \alpha < \infty$, has the property $y \to \infty$, $y' \to \infty$ as $x \to \infty$.

Let us show first that $y'(x) > 0$ for $x > \alpha$. In fact, let $x_1 > \alpha$ be the first zero of $y'(x)$ to the right of α. Then $y(x)>0$, $y'(x)>0$ in (α, x_1), and hence

$$y'''(x) = -2A(x)y'(x) - [A'(x) + b(x)]y(x) \geqq 0 .$$

From the last inequality we deduce that $y''(x) \geqq y''(\alpha)$, $y'(x) \geqq y'(\alpha) + y''(\alpha)\ (x-\alpha) > 0$ for $x > \alpha$, which implies $y'(x_1) \neq 0$ and $y(x) \to \infty$, $y'(x) \to \infty$ as $x \to \infty$.

The theorem is then proved. □

LEMMA 3.1. *Let $A(x) \leqq 0$, $A'(x) + b(x) \geqq 0$ and let $b(x)$ have the property (v) in (a, ∞). Let w be a solution of the differential equation (b) with $w(\alpha) = w'(\alpha) = 0$, $w''(\alpha) > 0$. Then $w(x) \to \infty$, $w'(x) \to \infty$ as $x \to \infty$.*

PROOF. Lemma 2.4 implies that $w(x) > 0$, $w'(x) > 0$ for $x > \alpha$.

From the integral identity (1.9) for w it follows that

$$w''(x) = -2A(x)w(x) + w''(\alpha) + \\ + \int_\alpha^x [A'(t) + b(t)]w(t)\, dt > 0$$

for $x > \alpha$. Thus $w''(x) \geqq w''(\alpha)$ for $x > \alpha$. Integrating this inequality, we get

$$w'(x) \geqq w''(\alpha)\ (x-\alpha)\ , \quad w(x) \geqq \frac{w''(\alpha)}{2}\ (x-\alpha)^2, \tag{3.1}$$

which implies the assertion. □

LEMMA 3.2. *Let the hypotheses of Lemma 3.1 be satisfied and, moreover, let*

$$\int_a^\infty t^2[A'(t) + b(t)]\, dt = \infty .$$

Then the solution w mentioned in Lemma 3.1 satisfies also $w''(x) + 2A(x)w(x) \to \infty$ as $x \to \infty$.

The proof of the lemma follows from the integral identity

$$w''+2Aw=w''(\alpha)+\int_\alpha^\infty (A'+b)w\,dt$$

and from the inequality (3.1). □

THEOREM 3.3. Let the hypotheses of Lemma 3.2 be satisfied in (a, ∞). Then there exists exactly one solution y of the differential equation (a) (up to linear dependence) with the following properties: $y(x)\neq 0$ for $x\in(a, \infty)$, y, y', y'' are monotonic functions of $x\in(a, \infty)$, $\operatorname{sgn} y=\operatorname{sgn} y''\neq\operatorname{sgn} y'$ for $x\in(a, \infty)$ and $y\to 0$, $y'\to 0$, $y''\to 0$ as $x\to\infty$.

PROOF. According to Theorems 2.10 and 2.11, there is at least one solution y of (a) with the following properties: $y(x)\neq 0$ for $x\in(a, \infty)$, y, y', y'' are monotonic functions of $x\in(a, \infty)$ and $\operatorname{sgn} y=\operatorname{sgn} y''\neq\operatorname{sgn} y'$, $a<x<\infty$, while $y'\to 0$, $y''\to 0$ as $x\to\infty$. We shall now show that $y\to 0$ as $x\to\infty$. For the sake of brevity assume that $y(x)>0$, $a<x<\infty$. Further, let y_1, y_2 be solutions of the differential equation (a) with

$$y_1(\alpha)=y_1'(\alpha)=0,\quad y_1''(\alpha)>0,\quad y_2(\alpha)=y_2''(\alpha)=0\,,$$
$$y_2'(\alpha)<0.$$

It is easy to verify that y_1, y_2, y is a fundamental set of solutions of (a) and that

$$\begin{vmatrix} y_1, & y_2, & y \\ y_1', & y_2', & y' \\ y_1'', & y_2'', & y'' \end{vmatrix} = -y_1''(\alpha)y_2'(\alpha)y(\alpha)>0\,.$$

The last equality implies the equation

$$wy''-w'y'+(w''+2Aw)y=-y_1''(\alpha)y_2'(\alpha)y(\alpha)>0\,,\quad (3.2)$$

where $w=y_1y_2'-y_1'y_2$ is a solution of the differential equation (b) with $w(\alpha)=w'(\alpha)=0$, $w''(\alpha)>0$, and hence $w(x)>0$ for $x>\alpha$. From (3.2) and Lemmas 3.1 and 3.2 it follows that $y\to 0$ as $x\to\infty$.

It remains to prove that there is only one solution y (up to linear dependence) of the differential equation (a) with the properties described above.

If the differential equation (a) has an oscillatory solution, uniqueness can be proved as in Theorem 2.13. Let the differential equation (a) have no oscillatory solution and let y, $\bar{y}$ be two independent solutions of (a), different from zero in (a, ∞) and converging monotonically to zero. Clearly, there exist c_1, c_2 and c_3 with $\bar{y} = c_1 y_1 + c_2 y_2 + c_3 y$, where y_1, y_2 have the above properties. Besides, we have

$$\bar{y} - y = c_1 y_1 + c_2 y_2 + (c_3 - 1) y \ . \tag{3.3}$$

By Lemma 2.6, there exist $x_1 > a$ such that the solution $c_1 y_1 + c_2 y_2 = Y$ satisfies $Y(x) Y'(x) \geqq 0$. Then (3.3) implies that $\lim_{x \to \infty} (\bar{y} - y) = 0$, contradicting the assumption that $\bar{y} \to 0$, $y \to 0$ as $x \to \infty$. This completes the proof. □

THEOREM 3.4. Let $b(x)$ have the property (v) in (a, ∞) and let $\int_a^\infty b \, dt$ diverge. Then the differential equation (a) has at least one solution y with no zeros in (a, ∞) and satisfying

$$\liminf_{a \leqq x < \infty} y(x) = 0 \ .$$

PROOF. We conclude from Theorem 1.13 that there exists a solution y of (a) with $y(x) \neq 0$ for $x \in (a, \infty)$ and satisfying the integral identity (1.13),

$$yy'' - \frac{1}{2} y'^2 + Ay^2 = \int_x^\infty by^2 \, dt \ .$$

This identity implies that $\int_\alpha^\infty by^2 \, dt$ converges, hence the assertion of the theorem follows. □

COROLLARY 3.1. Let $b(x) \geqq m > 0$, $a < x < \infty$, where m is a constant. Then there exists at least one solution $y(x) \neq 0$, $a < x < \infty$, of the differential equation (a) which belongs to class L^2 in $\langle \alpha, \infty)$, $a < \alpha < \infty$.

The integral identity (1.13) implies that

$$\int_\alpha^\infty y^2\,dt \leqq \frac{1}{m}\int_\alpha^\infty by^2\,dt < \infty\,,$$

whence the assertion follows.

COROLLARY 3.2. Assume that $b(x)$ has the property (v), $A(x)\leqq 0$ in (a, ∞) and that $\int_\alpha^\infty b\,dt$ diverges, $a<\alpha<\infty$. Then there exists at least one solution of the differential equation (a) $y(x)\neq 0$ in (a, ∞) having the following properties in (x_0, ∞):

$$\operatorname{sgn} y = \operatorname{sgn} y'' \neq \operatorname{sgn} y',$$

$$\lim_{x\to\infty} y = \lim_{x\to\infty} y' = \liminf_{\alpha\leqq x<\infty} y'' = 0.$$

PROOF. Let $y(x)\neq 0$ in (a, ∞) be a solution of (a) satisfying (1.13). We show that $y'(x)\neq 0$ for $x\in(a, \infty)$. Let $y(x)>0$ for $x\in(a, \infty)$. The integral identity (1.13) implies that $y''(x)>0$ for $x\in(a,\infty)$. If it were true that $y'(\bar{x})=0$ for $a<\bar{x}<\infty$, then y would attain a minimum at $\bar{x}$ and, for $x>\bar{x}$, we would have $y'(x)>0$. In virtue of Theorem 3.4, this is impossible. Therefore, $y'(x)<0$ for $x\in(a, \infty)$ and $\operatorname{sgn} y = \operatorname{sgn} y''\neq \operatorname{sgn} y'$ for $x\in(a, \infty)$. It obviously follows that $\lim_{x\to\infty} y(x) = \lim_{x\to\infty} y'(x)=0$, otherwise y would have a zero. Since $y'(x)>0$, $y'(x)$ is an increasing function of $x\in(a, \infty)$. Therefore, $\liminf_{\alpha\leqq x<\infty} y''(x)=0$, otherwise $y'(x)$ would have a zero, which is impossible. This completes the proof.□

THEOREM 3.5. Let $b(x)$ have the property (v), $A(x)\leqq 0$ and $A'(x)+b(x)\geqq 0$ for $x\in(a, \infty)$ and assume that $\int_\alpha^\infty b\,dt$ diverges, $a<\alpha<\infty$. Then there exists exactly one solution y of the differential equation (a) with the following properties: $y(x)\neq 0$, y, y', y'' are monotonic functions and $\operatorname{sgn} y = \operatorname{sgn} y''\neq \operatorname{sgn} y'$ for $x\in(a,\infty)$, while $y\to 0$, $y'\to 0$, $y''\to 0$ as $x\to\infty$.

PROOF. If the differential equation (a) is oscillatory in (a, ∞), then by Corollary 3.2 and Theorem 2.13 there exists exactly one solution with the specified properties.

If the differential equation does not oscillate in (a, ∞), by Corollary 3.2 there exists at least one solution $y(x)$ with $y(x) \neq 0$ and such that y, y' are monotonic functions in (a, ∞), $\operatorname{sgn} y = \operatorname{sgn} y'' \neq \operatorname{sgn} y'$, $y \to 0$, $y' \to 0$ as $x \to \infty$ and $\liminf_{a \leqq x < \infty} y'' = 0$. We show that also y'' is a monotonic function. If $y(x) > 0$ for $x \in (a, \infty)$, then $y'(x) < 0$ for $x \in (a, \infty)$ and the differential equation (a) implies that $y'''(x) \leqq 0$ for $x \in (a, \infty)$. Therefore, $y''(x)$ is a decreasing function of $y \in (a, \infty)$.

The uniqueness of y with the specified properties can be proved as in Theorem 3.3, so the theorem is proved. □

The following theorem is a generalization of Theorem 3.1.

THEOREM 3.6. Let $A(x) \geqq 0$, $A'(x) + b(x) \geqq d > 0$, $b(x) - A'(x) \geqq 0$ for $x \in (a, \infty)$. Then every solution of the differential equation (a) is oscillatory in (a, ∞), except a solution y (unique up to linear dependence), which satisfies $y(x) \in L^2\langle \alpha, \infty)$, $y \to 0$, $y' \to 0$ as $x \to \infty$, $a < \alpha < \infty$.

PROOF: In view of Theorem 2.16, the differential equation (a) is oscillatory in (a, ∞), i.e. each of its solutions having a zero oscillates in (a, ∞).

Let y be any non-oscillatory solution of (a). Then $y(x) \neq 0$ for $x \in (a, \infty)$. Assume that $y(x) > 0$ for $x \in (a, \infty)$.

It follows from the integral identity (1.7) for the solution y that

$$y'' + 2Ay = y''(\alpha_1) + 2A(\alpha_1)y(\alpha_1) - \int_{\alpha_1}^{x} (b - A')\, y \, dt <$$
$$< y''(\alpha_1) + 2A(\alpha_1)y(\alpha_1), \quad a < \alpha_1 < \infty .$$

Thus $y''(x) + 2A(x)y(x)$ is bounded in $\langle \alpha_1, \infty)$, and since $2A(x)y(x) \geqq 0$ for $x \geqq \alpha_1$, $y''(x)$ is bounded from above for $x \geqq \alpha_1$. Therefore, there exists a positive constant k such that

$$2y''(x) + 2A(x)y(x) \leqq k \quad \text{for} \quad x \geqq \alpha_1 . \tag{3.4}$$

Due to (a) being oscillatory in (a, ∞), Theorem 2.17 implies that $F(y(x))>0$ for $x \geqq \alpha_2$, $a<\alpha_2<\infty$, i.e.

$$F(y(x)) = y(x)y'(x) - \frac{1}{2}y'^2(x) + A(x)y^2(x) =$$

$$= F(y(\alpha_2)) - \int_{\alpha_2}^{x} b(t)y^2(t)\,dt > 0 \tag{3.5}$$

for $x \geqq \alpha_2$.
Therefore,

$$d\int_{\alpha_2}^{\infty} y^2(t)\,dt \leqq \int_{\alpha_2}^{\infty} [A'(t)+b(t)]y^2(t)\,dt \leqq$$

$$\leqq \int_{\alpha_2}^{\infty} b(t)y^2(t)\,dt \leqq F(y(\alpha_2)) < \infty\,,$$

implytig that $y(x) \in L^2\langle \alpha_2, \infty)$.
From (3.4) and (3.5) it follows that there exists $\alpha \geqq \alpha_i$, $i=1, 2$, such that

$$y'^2(x) \leqq [2y''(x) + 2A(x)y(x)]y(x) < ky(x) \quad \text{for} \quad x > \alpha\,. \tag{3.6}$$

If y does not tend to zero as $x \to \infty$, then for a given $\varepsilon > 0$ there exist arbitrarily great numbers x for which $y(x) > 2\varepsilon$. On the other hand, $y(x) \in L^2\langle \alpha, \infty)$, therefore there exist arbitrarily great x with $y(x) < \varepsilon$. This implies that there are number sequences $\{x_n\}_{n=1}^{\infty}$, $\{x_n^*\}_{n=1}^{\infty}$ such that $x_n < x_n^* < x_{n+1}$ while $\lim_{n\to\infty} x_n = \lim_{n\to\infty} x_n^* = \infty$ and moreover $y(x_n) < \infty$, $y(x_n^*) > 2\varepsilon$.
However, y being continuous, there exist sequences $\{z_n\}_{n=1}^{\infty}$, $\{z_n^*\}_{n=1}^{\infty}$ such that $x_n < z_n < z_n^* < x_n^*$, while

$$y(z_n) = \varepsilon\,, \quad y(z_n^*) = 2\varepsilon \tag{3.7}$$

and

$$\varepsilon < y(x) < 2\varepsilon \quad \text{for } (z_n, z_n^*)\,. \tag{3.8}$$

Since $y(x) > 0$ for $x \geqq \alpha$ and no two intervals (z_n, z_n^*) have any points in common, it follows from (3.8) that

$$\sum_{n=1}^{\infty} (z_n^* - z_n)\varepsilon^2 \leqq \sum_{n=1}^{\infty} \int_{z_n}^{z_n^*} y^2(t)\, dt \leqq \int_a^{\infty} y^2(t)\, dt\,,$$

therefore

$$\lim_{n\to\infty} (z_n^* - z_n) = 0\,. \tag{3.9}$$

The mean value theorem implies the existence of a sequence of points ξ_n, $z_n < \xi_n < z_n^*$ such that

$$y'(\xi_n) = \frac{y(z_n^*) - y(z_n)}{z_n^* - z_n} = \frac{\varepsilon}{z_n^* - z_n}\,. \tag{3.10}$$

Therefore, (3.9) implies that $\lim_{n\to\infty} y'(\xi_n) = \infty$, but from (3.6) and (3.8) it follows that

$$y'^2(\xi_n) < 2k\varepsilon\,. \tag{3.11}$$

Thus the assumption $\lim_{x\to\infty} y(x) \neq 0$ gives rise to a contradiction so

$$\lim_{x\to\infty} y(x) = 0\,, \tag{3.12}$$

and (3.6) implies also that

$$\lim_{x\to\infty} y'(x) = 0\,. \tag{3.13}$$

Now, let us prove the uniqueness of the solution without zeros. By Theorem 1.13, there exists at least one solution y with no zeros in (a, ∞). Suppose that $z(x)$ is another solution without zeros in (a, ∞), independent of y. Choose c so that $z(\alpha) - cy(\alpha) = 0$. Obviously, the solution $w(x) = z(x) - cy(x)$ is oscillatory in (a, ∞), because it vanishes at α. Let $\bar{x} > \alpha$ be a zero of $w(x)$. Evidently,

$$F(w(\alpha)) = -\frac{1}{2}\, w'^2(\alpha) \leqq 0$$

and

$$F(w(x)) = -\frac{1}{2}\, w^2(\bar{x}) = F(w(\alpha)) - \int_a^x b(t) w^2(t)\, dt <$$

$$<F(w(\alpha))\leqq 0\,,$$

that is

$$-F(w(x))=\frac{1}{2}\,w'^2(\bar{x})>-F(w(\alpha))\geqq 0\,.$$

Since $w(x)$ (being oscillatory) has zeros in any neighbourhood of ∞, we have

$$\limsup_{x\to\infty}\frac{1}{\sqrt{2}}\,|w'(x)|>\sqrt{-F(w(x))}>0\,.$$

On the other hand, since y and z are non-oscillatory solutions of the differential equation (a), (3.13) implies that

$$\lim_{x\to\infty} w'(x)=\lim_{x\to\infty}\,[z'(x)-cy'(x)]=0\,,$$

which is a contradiction. The theorem is thus proved. □

THEOREM 3.7. Let $A(x)\leqq 0$, $A'(x)\leqq 0$, $A'(x)+b(x)\geqq 0$ and let $b(x)$ have the property (v) in (a,∞). Moreover, let $\lim_{x\to\infty} A(x)\neq 0$. Then there exists exactly one solution y (up to linear dependence) of the differential equation (a) with the following properties: $\operatorname{sgn} y=\operatorname{sgn} y''\neq\operatorname{sgn} y'$, $x\in(a,\infty)$, y, y', y'' are monotonic functions of $x\in(a,\infty)$, $y\to 0$, $y'\to 0$, $y''\to 0$ for $x\to\infty$ and, finally,

$$\lim_{x\to\infty} A(x)y^2(x)=0\,. \tag{3.14}$$

PROOF. By Theorems 2.10 and 2.11, there is at least one solution y of the differential equation (a) such that $\operatorname{sgn} y=\operatorname{sgn} y''\neq\operatorname{sgn} y'$ for $x\in(a,\infty)$, $y'\to 0$, $y''\to 0$ as $x\to\infty$ and $\lim_{x\to\infty} y(x)$ exists and is finite.

From the proof of Theorem 2.10 it follows that one of the solutions having these properties is the solution $y(x)>0$ satisfying the identity (1.13) (Theorem 1.13). That identity (1.13) implies that

$$F(y(x)) = y(x)y''(x) - \frac{1}{2}y'^2(x) + A(x)y^2(x) \geqq 0$$

for $x \in (a, \infty)$.

The function $F(y(x))$ is decreasing in (a, ∞), hence $\lim_{x\to\infty} F(y(x))$ exists and is non-negative. Since $y' \to 0$, $y'' \to 0$ as $x \to \infty$, we have $\lim_{x\to\infty} F(y(x)) = A(x)y^2(x) \geqq 0$.

Because $A(x) \leqq 0$ and $\lim_{x\to\infty} A(x) \neq 0$, it follows necessarily that

$$\lim_{x\to\infty} A(x)y^2(x) = 0\,, \quad \lim_{x\to\infty} y(x) = 0\,.$$

Uniqueness is established as in Theorem 3.3, and the theorem is proved. □

REMARK 3.1. It can be proved (Švec [137]) that if $A(x) \leqq 0$, $A'(x) + b(x) \geqq 0$, $[A'(x) + b(x)]' \geqq 0$ for $x \in (a, \infty)$ and $A'(x) + b(x) \not\equiv 0$ in any subinterval of (a, ∞), then the assertion of Theorem 3.7 is valid except for (3.14), instead of which we have $\lim_{x\to\infty} [A'(x) + b(x)]y^2(x) = 0$.

THEOREM 3.8. Let $b(x)$ have the property (v) in (a, ∞) and, moreover, let $A(x) \leqq 0$ for $x \in (a, \infty)$. Then there exists at least one solution z of the differential equation (b) for $x \in (a, \infty)$ with $z(x) > 0$, $z'(x) > 0$, $z''(x) > 0$ for $x \in (a, \infty)$ and $\lim_{x\to\infty} z(x) = \infty$.

PROOF. From Remark 1.5 (assuming $b(x)$ has the property (v) in (a, ∞)) it follows that there is at least one solution of (b), $z(x) > 0$ in (a, ∞) satisfying the following integral identity (analogous to (1.13))

$$zz'' - \frac{1}{2}z'^2 + Az^2 = \int_a^x b(t)z^2(t)\,\mathrm{d}t + k, \quad k \geqq 0\,. \tag{3.15}$$

The solution z is the limit of a sequence of solutions $\{z_n\}_{n=1}^{\infty}$ with

$z_n(x_n)=z_n'(x_n)=0$, $z_n''(x_n)>0$, where $\{x_n\}_{n=1}^{\infty}$ is a point sequence tending to a.

By (3.15), $z''(x)>0$ for $x\in(a,\infty)$, hence $z'(x)$ is an increasing function of $x\in(a,\infty)$. From Lemma 2.4 we deduce that every solution z_n has $z_n'(x)>0$ for $x>x_n$. In view of $\lim_{n\to\infty} z_n'(x)=z'(x)\geqq 0$ and $z''(x)>0$ for $x\in(a,\infty)$, it follows that $z'(x)>0$ for $x\in(a,\infty)$. Then evidently $\lim_{x\to\infty} z(x)=\infty$, so the theorem is proved. □

THEOREM 3.9. *Let $A(x)\leqq 0$, $A'(x)-b(x)\leqq 0$ for $x\in(a,\infty)$ and let $b(x)$ have the property (v) in (a,∞). Then there is at least one solution z of the differential equation (b) satisfying $z(x)>0$, $z'(x)>0$, $z''(x)>0$ for $x\in(a,\infty)$, while $\lim_{x\to\infty} z(x)=\lim_{x\to\infty} z'(x)=\infty$.*

The proof follows from Theorem 3.8 and Lemma 2.12. □

REMARK 3.2. Theorem 3.9 can be proved even without the assumption that $b(x)$ has the property (v) in (a,∞). To do so, one makes use of the fact that also in this case the differential equation (b) is in class II (Švec [137]).

THEOREM 3.10. *Let $A(x)\geqq 0$, $A'(x)-b(x)\leqq 0$ for $x\in(a,\infty)$, let $b(x)$ have the property (v) in (a,∞), and let $\int_{x_0}^{\infty} b(t)\,dt=\infty$, $a<x_0<\infty$. Also, let $f(x)$ be a non-negative function having a continuous third derivative in (a,∞) satisfying $f'''+2Af'+(A'-b)f\leqq 0$ for $x\in(a,\infty)$. Then, given any solution z of the differential equation (b) with no zeros in (a,∞), $x_0\leqq \alpha<\infty$, there exist numbers $k>0$, $\xi\geqq\alpha$ such that $|z|-kf>0$ for $x\in(\xi,\infty)$.*

PROOF. Let $z(x)$ be a solution of (b) with $z_1(x_0)=f(x_0)$, $z_1'(x_0)=f'(x_0)$, $z_1''(x_0)=f''(x_0)$. The difference $u(x)=z_1(x)-f(x)$ satisfies the following differential equation:

$$u'''+2Au'+(A'-b)u=-(f'''+2Af'+(A'-b)f)\,.$$

Using the method of variation of constants, it is easy to see that $u(x)$ can be written in the form (cf. Lemma 2.3)

$$u(x) = -\int_{x_0}^{x} [f'''(t) + 2A(t)f'(t) + \\ +(A'(t) - b(t))f(t)]\,W(x, t)\,dt\,, \tag{3.16}$$

where $W(x, t)$ for a fixed t is a solution of the differential equation (b) having a double zero at t and no other zero for $x > t$. Evidently, $W(x, t) \geqq 0$ for $x \geqq t$, because $W''_x(t, t) = 1$. From (3.16) it follows that $u(x) \geqq 0$ for $x > x_0$, that is, $z_1(x) \geqq f(x)$ for $x > x_0$.

Now let $z(x)$ be any solution of (b) with no zeros in (α, ∞), $x_0 \leqq \alpha < \infty$, and let $z(x) > 0$ for $x \in (\alpha, \infty)$.

First of all, we show that $z'(x) \leqq 0$ for $x \in (\alpha, \infty)$ is impossible. Assuming that $z'(x) \leqq 0$, the differential equation (b) and the hypotheses imply that $z'''(x) \geqq 0$ for $\alpha < x < \infty$, hence $z''(x)$ is a non-decreasing function of $x \in (\alpha, \infty)$, and therefore $z''(x) \to \beta$ as $x \to \infty$. If $\beta > 0$, we have $z'(x) \geqq 0$ starting from some x, which is a contradiction. If $\beta < 0$, then necessarily $\lim_{x\to\infty} z'(x) = -\infty$ and also $z(x) \to -\infty$, which again gives rise to a contradiction. If $\beta = 0$, then $z''(x) \leqq 0$, and hence $z'(x) \leqq 0$ is a non-increasing function of x, implying again that $z(x) \to -\infty$ as $x \to \infty$, which is not possible. Therefore, $z'(x) \geqq 0$ beginning at some x or $z''(x)$ changes its sign infinitely many times. In the former case we have $z(x) \geqq c > 0$ starting from some $x = \bar{x}$. Therefore,

$$\int_{\bar{x}}^{x} bz^2\,dt \geqq c^2 \int_{\bar{x}}^{x} b\,dt \to \infty \quad \text{as } x \to \infty\,.$$

The integral identity (1.8) implies the existence of some $\xi_1 \in (\alpha, \infty)$ at which

$$z(\xi_1)z''(\xi_1) - \frac{1}{2}z'^2(\xi_1) + A(\xi_1)z^2(\xi_1) > 0\,.$$

In the latter case, z attains a local minimum infinitely many times. Let ξ_2 be one of the points at which $z(x)$ attains such a minimum. Obviously, $z(\xi_2) > 0$, $z'(\xi_2) = 0$, $z''(\xi_2) > 0$, and also

$$z(\xi_2)z''(\xi_2)-\frac{1}{2}z'^2(\xi_2)+A(\xi_2)z^2(\xi_2)>0\ .$$

So, let $\xi>\alpha$ be a point at which

$$z(\xi)z''(\xi)-\frac{1}{2}z'^2(\xi)+A(\xi)z^2(\xi)>0\ .$$

Write $z_2(x)\overset{\mathrm{d}}{=}z(x)-kz_1(x)$ and choose $k>0$ so small that

$$z_2(\xi)>0,\qquad z_2(\xi)z_2''(\xi)-\frac{1}{2}z'^2(\xi)+A(\xi)z^2(\xi)>0\ ,$$

(this is evidently possible).

The integral identity (1.8) for the solution z_2 reads

$$z_2z_2''-\frac{1}{2}z_2'^2+Az_2^2-\int_\xi^x bz_2^2\,\mathrm{d}t=$$
$$=z_2(\xi)z_2'(\xi)-\frac{1}{2}z_2'^2(\xi)+A(\xi)z_2^2(\xi)>0\ .$$

Assuming that $z_2(\bar{x})=0$ for $\bar{x}>\xi$, we get a contradiction. Thus $z_2(x)>0$ for $x>\xi$, and therefore $z(x)>kz_1(x)\geqq kf(x)$ for $x>\xi$. The theorem is proved. □

COROLLARY 3.3. Let $A(x)\geqq 0$, $2A(x)+1+A'(x)-b(x)\leqq 0$ for $x\in(\alpha,\infty)$. Then non-oscillatory solutions of the differential equation (b) diverge to $\pm\infty$ more rapidly than $k\,\mathrm{e}^x$, where k is a suitable constant.

Evidently, the hypotheses of Theorem 3.10 hold true in Corollary 3.3 since $\mathrm{e}^x(1+2A+A'-b)\leqq 0$, and also $2A+1+A'\leqq b$ implies that $\int_a^\infty b\,\mathrm{d}t$ diverges.

THEOREM 3.11. Let $A(x)\leqq 0$, $b(x)-|A'(x)|\geqq k>0$ for $x\in(a,\infty)$ and let the differential equation (a) be oscillatory in (a,∞). Then every solution of the differential equation (b) having no zeros starting from some x has the following property:

$$\lim_{x\to\infty} z(x) = \lim_{x\to\infty} z'(x) = \lim_{x\to\infty} z''(x) = \pm\infty .$$

PROOF. Let $z(x)$ be a solution of (b) with $z(x)>0$ for $x\in(\alpha, \infty)$, $a<\alpha<\infty$. According to Lemma 1.1, there are two independent solutions y_1, y_2 of (a) such that $z(x)=y_1(x)y_2'(x)-y_1'(x)y_2(x)\neq 0$ for $x>\alpha$. By the hypotheses, the band of solutions $y=c_1y_1+c_2y_2$ is oscillatory in (α, ∞) and satisfies in (α, ∞) the differential equation of the form (c) with $w=z(x)\neq 0$ for $x>\alpha$. Oscillatoricity of the band in (α, ∞) implies that at some $\bar{x}>\alpha$ we have $z''(\bar{x})+2A(\bar{x})z(\bar{x})>0$, that is, $z''(\bar{x})>0$, owing to the assumption $A(x)\leqq 0$ for $a<x<\infty$. It follows from the integral identity (1.9) for the solution z that for $x>\bar{x}$ we have

$$z''(x) = z''(\bar{x}) + 2A(\bar{x})z(\bar{x}) - 2A(x)z(x) + \\ + \int_{\bar{x}}^{x} [A'(t)+b(t)]z(t)\,\mathrm{d}t > 0 ;$$

hence $z'(x)$ is an increasing function of $x>\bar{x}$. We have $z''(x)\geqq z''(\bar{x})+2A(\bar{x})z(\bar{x})=k>0$, $z'(x)\geqq k(x-\bar{x}) + z'(\bar{x})$, therefore $z'(x)\to\infty$ as $x\to\infty$ and also $z(x)\to\infty$ as $x\to\infty$. We still have to prove that $z''(x)\to\infty$ as $x\to\infty$. The differential equation (b) implies that there is $x_1>\bar{x}$ such that $z'''(x)>k>0$ for $x>x_1$, whence the assertion follows. This completes the proof. □

REMARK 3.3. The hypotheses of Theorem 3.11 are identical with those of Theorem 2.15.

REMARK 3.4. The type of theorems like Theorem 3.11 comprises also Theorem 2.47.

LEMMA 3.3. Let $A(x)\leqq 0$, $A'(x)+b(x)\geqq 0$ for $x\in(a, \infty)$ and let $b(x)$ have the property (v) in (a, ∞). Also, let the differential equation (a) be oscillatory in (a, ∞). Any non-oscillatory solution u of (a) then satisfies $\lim_{x\to\infty} xu'(x)=0$.

PROOF. By Theorem 2.13, there is (up to linear dependence) exactly one non-oscillatory solution $u(x)$ of (a) with the following properties:

$u(x) \neq 0$, $\operatorname{sgn} u(x) = \operatorname{sgn} u''(x) \neq \operatorname{sgn} u'(x)$ for $x \in (a, \infty)$, u, u', u'' are monotonic functions of $x \in (a, \infty)$, and $\lim_{x\to\infty} u'(x) = \lim_{x\to\infty} u''(x) = 0$, $\lim_{x\to\infty} u(x) = c$, where c is a finite constant.

Let $u(x) < 0$ for $x \in (a, \infty)$, then $u'(x) > 0$, $u''(x) < 0$ for $x \in (a, \infty)$ and $c < 0$. Clearly, $\int_a^\infty u'(t)\,dt < \infty$. Let $\varepsilon > 0$. There is a constant $N > 0$ such that $\int_N^x u'(t)\,dt < \varepsilon$ for $x > N$, hence

$$\varepsilon > \int_N^x u'(t)\,dt = u'(\xi)\,(x - N) \quad \text{for} \quad N < \xi < x\,.$$

On the other hand, $u''(x) < 0$, therefore $u'(\xi)(x - N) \geqq u'(x) \,.\, (x - N) > u'(x)x - \varepsilon$ for large x, because $u'(x) \to 0$. Thus $2\varepsilon > xu'(x)$ for great x, whence $\lim_{x\to\infty} xu'(x) = 0$. The lemma is proved. □

LEMMA 3.4. Let the hypotheses of Lemma 3.3 be satisfied and let $u(x)$ be a solution of the differential equation (a) with the properties listed in Lemma 3.3. Then

$$\left| \int_a^\infty xu''(x)\,dx \right| < \infty\,.$$

PROOF. Let $u(x) > 0$, $u'(x) < 0$ and $u''(x) > 0$. Integration by parts yields

$$\int_a^\infty tu''(t)\,dt = xu'(x) - \alpha u'(\alpha) - u(x) + u(\alpha)\,.$$

Therefore $\int_a^\infty xu''(x)\,dx < \infty$, because $\lim_{x\to\infty} xu'(x) = 0$ and $\lim_{x\to\infty} u(x) = c \neq \pm\infty$, $a < \alpha < \infty$. □

LEMMA 3.5. Let the hypotheses of Lemma 3.3 hold true and let $u(x)$ be as in Lemma 3.3. Then $\lim_{x\to\infty} x^2 u'(x) = 0$.

PROOF. Let $u(x)>0$, $u'(x)<0$, $u''(x)>0$ for $x\in(a,\infty)$. Since $\int_{\alpha}^{\infty} xu''(x)\,dx<\infty$, $a<\alpha<\infty$, for any given $\varepsilon>0$ there exists $N>0$ such that for every $x>N$ we have

$$\varepsilon>\int_N^x tu''(t)\,dt=u''(\xi)\int_N^x t\,dt$$

for suitable $N<\xi<x$.

Since $u'''(x)<0$, which follows from the hypotheses and from the equation (a), this implies that

$$u''(\xi)\int_N^x t\,dt\geqq\frac{u'''(x)}{2}(x^2-N^2)\geqq\frac{u''(x)}{2}x^2-\frac{\varepsilon}{2}$$

for great x, in view of $\lim_{x\to\infty} u''(x)=0$. Therefore, $3\varepsilon>x^2u''(x)$ for all great x. Thus, $\lim_{x\to\infty} x^2u''(x)=0$. □

By Lemma 1.1 applied to the solution u of (a), there exist two independent solutions z_1 and z_2 of the differential equation (b), satisfying $u(x)=z(x)z_2'(x) - z_1'(x)z_2(x)\neq 0$ for $x\in(a,\infty)$. The band of solutions $z=c_1z_1+c_2z_2$ satisfies the second order differential equation

$$uz''-u'z'+(u''+2Au)z=0\,,$$

which can be reduced to the self-adjoint form

$$\left[\frac{1}{u}z'\right]'+\frac{u''+2Au}{u^2}z=0\,. \tag{3.17}$$

LEMMA 3.6. Let the hypotheses of Lemma 3.3 be satisfied and let $u(x)>0$, $u'(x)<0$, $u''(x)>0$ be a non-oscillatory solution having the same properties as in Lemma 3.3. Then there exist two independent oscillatory solutions z_1 and z_2 of the differential equation (b) in (a,∞) which satisfy the differential equation (3.17).

PROOF. It is sufficient to show that there exist two independent oscillatory solutions z_1 and z_2 of (b) for which $z_1z_2'-z_1'z_2=ku(x)$, $k\neq 0$ being a constant.

Let y_1, y_2 be independent solutions of (a) with $y_1(\alpha)=y_1'(\alpha)=0$, $y_1''(\alpha)=1$, $y_2(\alpha)=y_2''(\alpha)=0$, $y_2'(\alpha)=1$. Let $z_1(x)=y_1(x)u'(x)-y_1'(x)u(x)$, $z_2(x)=y_2(x)u'(x)-y_2'(x)u(x)$. The solutions z_1, z_2 thus defined have the properties claimed for them, because

$$\begin{vmatrix} z_2, & z_1 \\ z_2', & z_1' \end{vmatrix} = u \begin{vmatrix} u, & y_2, & y_1 \\ u', & y_2', & y_1' \\ u'', & y_2'', & y_1'' \end{vmatrix} = u \begin{vmatrix} u(\alpha), & 0, & 0 \\ u'(\alpha), & 1, & 0 \\ u''(\alpha), & 0, & 1 \end{vmatrix} = u(\alpha)u(x)\,.$$

Thus, the lemma is proved. □

THEOREM 3.12. Let the hypotheses of Lemma 3.3 be satisfied and let $u(x)\neq 0$ for $x\in(a,\infty)$. Then $\lim_{x\to\infty} u(x)=0$.

PROOF. Assume the contrary, i.e. that $\lim_{x\to\infty} u(x)=k<\infty$, $k>0$ constant, whenever $u(x)>0$, which can be done without loss of generality. For the sake of brevity, put $k=1$. For large x we have $u(x)<2$, hence $\frac{1}{u(x)}>\frac{1}{2}$. Also,

$$u''(x)\geqq\frac{u''(x)}{u^2(x)}\geqq\frac{u''(x)}{u^2(x)}+\frac{2A(x)u(x)}{u^2(x)}\,.$$

Since by Lemma 3.6 the differential equation (3.17) is oscillatory in (a,∞), it follows from the Sturm comparison theorem (Sansone [124]) for second order equations that the differential equation

$$\left(\frac{v'}{2}\right)+u''(x)v=0 \tag{3.18}$$

is also oscillatory.

Putting $v=\sqrt{x}\,z$ for $x>0$ in (3.18), we get

$$(xz')'+\left(2x^2u''(x)-\frac{1}{4}\right)\frac{1}{x}\,z=0\,. \tag{3.19}$$

As Lemma 3.5 implies that $\lim_{x\to\infty} x^2u''(x)=0$, the expression $(2x^2u''(x)-1/2)$ is negative for great x, and therefore the equation (3.18) is non-oscillatory in (a, ∞). The contradiction we have obtained implies $\lim_{x\to\infty} u(x)=0$, so the theorem is proved. □

REMARK 3.5. Theorem 3.12 is a generalization of Theorem 2.13. Theorem 3.12 can be proved even without the assumption that $b(x)$ has the property (v) in (a, ∞), because the differential equation (a) belongs to class I (Jones [71]).

2. Asymptotic Properties of Oscillatory Solutions of the Differential Equation (b)

THEOREM 3.13. Let $A(x)\geqq 0$, $b(x)\geqq 0$ for $x\in(a, \infty)$ and let

$$\infty > M = \limsup_{x\to\infty} \frac{2A(x)}{\sqrt{x}},$$

$$m = \limsup_{x\to\infty} \sqrt{x}\,[A'(x)-b(x)]<0 .$$

Then every non-trivial solution of the differential equation (b) is either oscillatory in (a, ∞) or diverges to $\pm\infty$ more rapidly than any positive power of x. A solution z is oscillatory in (a, ∞) if and only if $zz''-1/2z'^2+Az^2<0$. If $b(x)\geqq d>0$, then every oscillatory solution of the differential equation (b) is in class L^2.

PROOF. Every non-trivial solution of (b) either

a) vanishes infinitely many times in (a, ∞) or

b) is different from zero, beginning at some x.

a) In this case, z must be oscillatory in (x_0, ∞), $x_0>a$, for the contrary would imply that there is a sequence $\{x_n\}_{n=1}^{\infty}$ of points with $x_n\to\infty$ as $n\to\infty$ and $z(x_n)=z'(x_n)=0$. From the integral identity (1.8) for z it follows that $\int_{x_{n-1}}^{x_n} b(t)z^2(t)\,dt=0$, thus $b(x)\equiv 0$ for $x_{n-1}\leqq x\leqq x_n$, and as $x_n\to\infty$, $b(x)\equiv 0$ for $x\geqq x_1$. Therefore,

$$A(x)=A(\xi)+\int_{\xi}^{x}[A'(t)-b(t)]\,dt\leqq A(\xi)+$$

$$+\frac{1}{4}\int_{\xi}^{x}\frac{2m}{\sqrt{t}}\,dt\to-\infty$$

as $x\to\infty$, which contradicts the assumption that $A(x)\geqq 0$.

Suppose that $b(x)\geqq d>0$ and let $x_1<x_2<x_3<\ldots$ be the zeros of the solution z. The integral identity (1.8) for z implies that

$$zz''-\frac{1}{2}z'^2+Az^2-\int_{x_1}^{x}bz^2\,dt=-\frac{1}{2}z'^2(x_1)\,.$$

If $x=x_n$, we get

$$\int_{x_1}^{x_n}bz^2\,dt=\frac{1}{2}[z'^2(x_1)-z'^2(x_n)]\leqq\frac{1}{2}z'^2(x_1)\,.$$

It follows that

$$\int_{x_1}^{x_n}z^2(t)\,dt\leqq\frac{1}{d}\int_{x_1}^{x_n}b(t)z^2(t)\,dt\leqq\frac{1}{d}z'^2(x_1)\,,$$

proving that $z(x)$ is in class L^2.

b) Let $|z(x)|>0$, beginning at some x. Assume $z(x)>0$. We begin with showing that there exists a solution of (b) which diverges to ∞. To this end, construct the differential equation

$$y'''+P(x)y'+Q(x)y=0\,, \tag{3.20}$$

where

$$P(x)=M_1\sqrt{x}-\frac{(s-1)(s-2)}{x^2}\,,\quad Q(x)=\frac{m_1}{\sqrt{x}}\,,\quad x>0\,.$$

Also, assume that $0>m_1>m$, $M_1>M$, $s=-(m_1/M_1)>0$, m_1, M_1 being constants.

One of the solutions of (3.20) is $y_1(x)=x^s$. It is easy to see that, starting from some $x_0>0$, the following inequalities hold:

$$P(x)>2A(x)\geqq 0\,,\quad 0>Q(x)>A'(x)-b(x)\,. \tag{3.21}$$

Let z_1 be a solution of the differential equation (b) satisfying, at x_0, the same initial conditions as y_1. Put $u(x)=z_1(x)-y_1(x)$. We have $u(x_0)=u'(x_0)=u''(x_0)=0$, thus

$$F(u(x_0)) = u(x_0)u''(x_0) - \frac{1}{2}u'^2(x_0) + A(x_0)u^2(x_0) = 0\,.$$

Subtracting (3.20) from (b), we get, for $x > x_0$,

$$u''' + 2A(x)u' + [A'(x) - b(x)]u =$$
$$= [Q(x) - A'(x) + b(x)]y_1(x) + [P(x) - 2A(x)]y_1'(x) > 0\,.$$

Clearly, we have also $u'''(x_0) > 0$. This fact, together with the above initial conditions for $u(x)$, implies that in some right neighbourhood of x_0 we have $u(x_0) > 0$. Now, if we apply Lemma 2.3 to the above homogeneous equation and bear in mind that $W(x, t)$ is a solution of (b) having a double zero at t and satisfying $W(x, t) \geqq 0$ for $x \geqq t$ (this follows from (1.8) for the solution $W(x, t)$ at a fixed t), then we obtain the conclusion that $u(x) > 0$ for $x > x_0$. Therefore, $u(x) = z_1(x) - x^s$, and hence $z_1(x) > x^s$ and $z_1(x) \to \infty$ as $x \to \infty$. We have thus proved that there exists a solution $z_1(x)$ of (b) diverging to ∞ as $x \to \infty$.

Let us now turn back to the solution z. First we show that we cannot have $z'(x) \leqq 0$, beginning at some x. If this were the case, the equation (b) would imply that $z'''(x) > 0$ and so there would exist $\lim_{x\to\infty} z''(x) = c$.

The constant c would necessarily equal zero, because $c > 0$ would imply $z' \to \infty$ as $x \to \infty$, and $c < 0$ would imply $z' \to -\infty$, and hence $z \to \infty$ as $x \to \infty$, a contradiction. From the assumption that $z'''(x) > 0$ and $c = 0$ it follows that $z''(x) < 0$, and so $z'(x)$ would be a decreasing function of x, which would imply that $z \to \infty$ as $x \to \infty$. However, $z > 0$, and a contradiction arises.

Thus, either $z'(x) \geqq 0$, starting from some x, or $z'(x)$ changes its sign infinitely many times. In the former case, $z(x) \geqq k > 0$, beginning at some x. The hypotheses of the theorem imply that

$$\int_{\bar{x}}^{x} b(t)z^2(t)\,dt \geqq k^2 \int_{\bar{x}}^{x} b(t)\,dt = k_1 + k_2A(x) -$$
$$- k^2 \int_{\bar{x}}^{x} (A'(t) - b(t))\,dt \to \infty \quad \text{as} \quad x \to \infty\,.$$

From (1.8) it follows that $F(z(x_1)) > 0$ for some x_1. In the latter case, $z(x)$ attains a local minimum infinitely many times. If x_1 is a point at which z attains a minimum, then $z(x_1)>0$, $z'(x_1)=0$, $z''(x_1)>0$. Therefore again $F(z(x_1))>0$. Put $z_2(x)=z(x)-kz_1(x)$ and chose $k>0$ sufficiently small to make

$$z_2(x_1)>0\,, \quad F(z_2(x_1))>0\,. \tag{3.22}$$

(This is evidently possible.) From (3.22) and Lemma 2.3 it follows that $z_2(x)>0$ for $x>x_1$, that is, $z(x)>kz_1(x)$, thus $z(x)\to\infty$ as $x\to\infty$.

It remains to prove that a solution z of (b) is oscillatory in (x_0, ∞) if and only if $F(z(x))<0$ in (x_0, ∞).

Let $F(z(x))<0$ and assume that z is non-oscillatory in (x_0, ∞). The preceding reasoning shows that $|z(x)|\to\infty$ as $x\to\infty$, hence $z^2(x)>1$, beginning at some $\alpha>x_0$. Therefore

$$\int_\alpha^x b(t)z^2(t)\,\mathrm{d}t \geqq \int_\alpha^x b(t)\,\mathrm{d}t = c + A(x) -$$

$$-\int_\alpha^x [A'(t)-b(t)]\,\mathrm{d}t \to \infty \tag{3.23}$$

as $x\to\infty$.

On the other hand, formula (1.8) applied to z gives

$$\int_\alpha^x b(t)z^2(t)\,\mathrm{d}t = F(z(x)) - F(z(x_0)) \leqq -F(z(x_0))\,,$$

which contradicts (3.23).

Assuming that $F(z(x_0))\geqq 0$, we have $z(x)>0$ or $z(x)<0$ in some right neighbourhood of x_0. In the latter case, $F(-z(x_0))\geqq 0$. Lemma 2.3 implies in either case that $z(x)>0$ or $-z(x)>0$ for $x>\alpha$, thus z cannot oscillate in (x_0, ∞). The theorem is proved. □

REMARK 3.6. If the hypotheses of Theorem 3.13 are satisfied and, moreover, the function $b(x)$ has the property (v) in (x_0, ∞), then the zeros of any two independent oscillatory solutions of the differential equation (b) separate each other (Zlámal [146]).

THEOREM 3.14. *Let $A(x)\leqq 0$, $b(x)\geqq d>0$ for $x\in(a, \infty)$. Then every oscillatory solution z of the differential equation (b) belongs to class $L^2\langle\alpha, \infty)$, $a<\alpha<\infty$, and $\lim_{x\to\infty} z(x)=0$.*

PROOF. Since $z(x)$ is an oscillatory solution in (a, ∞), the integral identity (1.8) implies that

$$F(z(x)) = F(z(\alpha)) + \int_\alpha^x b(t)z^2(t)\, dt\,.$$

It is easy to verify that there exist arbitrarily large x with $F(z(x))$ non-positive, namely those x at which $z(x)=0$, because $F(z(x)) = -1/2z'^2(x)$ there.

Thus, for all $x \in \langle \alpha, \infty)$,

$$\int_\alpha^x z^2(t)\, dt \leqq \frac{1}{d}\int_\alpha^x b(t)z^2(t)\, dt \leqq -\frac{F(z(\alpha))}{d}\,.$$

Hence $\int_\alpha^x z^2(t)\, dt < \infty$, and therefore $z \in L^2\langle \alpha, \infty)$.

By Lemma 2.14, the derivative of every oscillatory solution of (b) is bounded in $\langle \alpha, \infty)$. This evidently implies that $\lim_{x\to\infty} z(x) = 0$. □

THEOREM 3.15. Let $A(x) \leqq d_1 < 0$, $b(x) \geqq 0$ for $x \in (a, \infty)$ and let the differential equation (b) have at least one oscillatory solution in (a, ∞). Then the following is true:

a) Every non-oscillatory solution diverges monotonically to $\pm\infty$ as $x \to \infty$.

b) If, moreover, $b(x) \geqq d_2 > 0$, then every oscillatory solution belongs to the space $L^2\langle \alpha, \infty)$, $a < \alpha < \infty$.

c) The zeros of any two independent oscillatory solutions of the differential equation (b) separate each other.

PROOF. a) First of all, we show that every solution z with $z(x_1) = z'(x_1) = 0$, $z''(x_1) \neq 0$, $a < x_1 < \infty$, diverges monotonically to ∞ or $-\infty$ as $x \to \infty$. Assume that $z''(x_1) > 0$. Then $z''(x) > 0$ in some right neighbourhood of x_1. We show that $z''(x) > 0$ for all $x > x_1$. Suppose that this is false and denote by ξ the minimum of all those $x > x_1$ at which $z''(x) \leqq 0$. We have $z''(\xi) = 0$ and (1.8) implies that

$$F(z(\xi)) = -\frac{1}{2}\, z'^2(\xi) + A(\xi)z^2(\xi) = \int_{x_1}^{\xi} b(t)z^2(t)\, dt\,,$$

which gives rise to a contradiction, because the left-hand side is negative while the right-hand side is non-negative. Therefore, $z''(x)>0$ for $x>x_1$, thus $z'(x)$ is increasing, and since $z'(x_1)=0$ we have $z'(x)\geqq k>0$, with k a suitable constant, beginning at some $x_2>x_1$. The last inequality implies that $z(x)$ is a monotonic function and $z(x)>k(x-x_2)+z(x_2)$, thus $z\to\infty$ as $x\to\infty$.

Let now z be any non-oscillatory solution of (b). Choose $x_1\in(a, \infty)$ sufficiently great in order that $z(x)\neq 0$ for $x>x_1$. Without loss of generality we may assume $z(x)>0$ for $x>x_1$. We proceed to show that the solution z is unbounded.

Suppose that there exists $M>0$ with $z(x)<M$ for $x>x_1$. Let z_1 be an oscillatory solution of (b) in (a, ∞). Choose two consecutive zeros of z_1, $\xi_2>\xi_1>x_1$, such that $z_1'(\xi_1)>0$, $z_1'(\xi_2)<0$. The function $w(x)=z(x)z_1'(x) - z'(x)z_1(x)$ is continuous in $\langle \xi_1, \xi_2\rangle$ and $w(\xi_1) = z(\xi_1)z_1'(\xi_1)>0$, $w(\xi_2)=z(\xi_2)z_1'(\xi_2)<0$. Therefore, $w(\xi)=0$ at some point ξ in the interior of (ξ_1, ξ_2). Thus, the system of equations

$$c_1z(\xi)+c_2z_1(\xi)=0,$$
$$c_1z'(\xi)+c_2z_1'(\xi)=0$$

has a non-zero solution c_1, c_2. The function $\bar{z}(x)=c_1z(x)+c_2z_1(x)$ is a solution of the differential equation (b) and has a double zero at ξ and, according to what has already been proved, $\lim_{x\to\infty} \bar{z}(x)=\infty$, which is a contradiction since $|\bar{z}(x_k)|=|c_1|z(x_k)<|c_1|M$ at zeros x_k of z_1. Therefore, every non-oscillatory solution z of the differential equation (b) is unbounded in (a, ∞). To complete the proof of a), it is sufficient to show that z cannot have infinitely many maxima and minima. We again prove this by contradiction. Denote by $\xi_k (k=1, 2, \ldots)$ the points at which z attains a maximum. Since $z(x)$ is an unbounded solution, given any $L>0$, there exist indices k with $z(\xi_k)>0$. It follows from (1.8) that

$$z(\xi_k)z''(\xi_k)+A(\xi_k)z^2(\xi_k)=\int_{x_1}^{\xi_k} b(t)z^2(t)\,\mathrm{d}t+F(z(x_1))\,.$$

The right-hand side is lower bounded, while on the left-hand side we have

$$z(\xi_k)z''(\xi_k)+A(\xi_k)z^2(\xi_k)<A(\xi_k)z^2(\xi_k)\leqq d_1z^2(\xi_k)\,,$$

thus the left-hand side attains arbitrarily great non-negative values, yielding a contradiction.

b) From the assumption that $b(x)\geqq d_2>0$ and from (1.8) it follows that $F(z(x))$ is an increasing function of x and is negative for every oscillatory solution in $\langle\alpha,\infty)$. In fact, at any zero x_k of the solution z we have $F(z(x_k))=-(1/2)z'^2(x_k)<0$. Therefore, by (1.8),

$$-F(z(\alpha))>\int_\alpha^x b(t)z^2(t)\,dt\geqq d_2\int_\alpha^x z^2(t)\,dt\,,$$

which implies the assertion b).

c) We have to show that whenever z_1 and z_2 are independent oscillatory solutions of (b), they have no zero in common and between any two neighbouring zeros $x_1<x_2$ of one solution (say, z_2) there is exactly one zero of the other solution (that is, z_1).

The proof will be carried out by contradiction. Suppose that no zero of z_1 lies in the interval (x_1, x_2). Without loss of generality we may assume that $z_1(x)>0$, $z_2(x)>0$ for $x\in(x_1, x_2)$, and $z_2(x_1)=z_2(x_2)=0$, $z_2'(x_1)>0$, $z_2'(x_2)<0$ (neither $z_2'(x_1)=0$ nor $z_2'(x_2)=0$ can happen because then z_2 would be non-oscillatory). The same reasoning as we used in the proof of the assertion a) shows that this assumption implies the existence of a solution of (b) having a double zero at $\xi\in\langle x_1, x_2\rangle$; therefore z then cannot oscillate and, by the assertion b) of the theorem we are now proving, we deduce that

$$\lim_{x\to\infty} z(x)=\pm\infty\,. \tag{3.24}$$

On the other hand, however, z is a linear combination of two oscillatory solutions z_1, z_2. We show that every oscillatory solution is bounded. Thus z, too, must be bounded.

In fact, let z be any oscillatory solution of (b). The function $F(z(x))$ is non-decreasing and negative, because if we denote by x_k the points of positive local maxima or negative minima of $z(x)$, we have $z(x_k)z''(x_k)\leqq 0$, $F(z(x_k))\leqq A(x_k)z^2(x_k)\leqq d_1z^2(x_k)\leqq 0$, thus the sequence $\{z(x_k)\}$ is bounded. This contradicts (3.24). Therefore, between x_1 and x_2 there is at least one zero of z_1. More than one zero of z_1

cannot exist in (x_1, x_2), because if there were at least two zeros, $\bar{x}_1 < \bar{x}_2$, there, then by repeating the argument with z_1, z_2 interchanged, it follows that at least one zero of z_2 would lie between $\bar{x}_1$ and $\bar{x}_2$, which is impossible.

It remains to show that z_1 and z_2 cannot have a common zero ξ. If this were the case, there would exist a constant $k \neq 0$ with $z_1'(\xi) = kz_2'(\xi)$ and the function $z(x) = z_1(x) - kz_2(x)$ would have a double zero at ξ, thus (3.24) would hold true and a contradiction would arise with z_1 and z_2 being bounded. This completes the proof. □

3. *Asymptotic Properties of All Solutions of the Differential Equation (a)*

Employing results by Ráb [114, 115] we are going to show that, under certain assumptions concerning the coefficients of the differential equation (a) in (a, ∞), the asymptotic properties of solutions of (a) resemble those of the self-adjoint differential equation (1.3) or of the second order differential equation (1.4).

Rewrite the differential equation (a) in the form

$$y''' + 2Ay' + A'y = -by .$$

By the method of variation of constants, as in Lemma 2.3, it is easy to verify that

$$y(x) = c_1u_1(x) + c_2u_2(x) + c_3u_3(x) - \\ - \int_\alpha^x \frac{b(t)y(t)}{W(t)} W(x, t)\,dt , \tag{3.25}$$

$a < \alpha < \infty$, where u_1, u_2, u_3 is a fundamental set of solutions of the differential equation (1.3), $W(x)$ is the wronskian of the solutions u_1, u_2, u_3, and

$$W(x, t) = \begin{vmatrix} u_1(x), & u_2(x), & u_3(x) \\ u_1(t), & u_2(t), & u_3(t) \\ u_1'(t), & u_2'(t), & u_3'(t) \end{vmatrix} ,$$

c_1, c_2, c_3 are constants chosen so that the solution y and the function $c_1u_1 + c_2u_2 + c_3u_3$ satisfy at α the same initial conditions. Clearly, for a fixed t, the function $W(x, t)$ solves the differential equation (1.3) and has a double zero at t. Let z_1, z_2 be a fundamental set of solutions of the differential equation (1.4). If we put $u_1 = z_1^2$, $u_2 = z_1 z_2$, $u_3 = z_2^2$, apply Section 2 in §1, and choose z_1, z_2 so that their wronskian be equal to 1, we calculate that $W(x) = 2$. Another short calculation, based on (3.25), yields

$$y(x) = c_1 z_1^2(x) + c_2 z_1(x) z_2(x) + c_3 z_2^2(x) - \\ -\frac{1}{2}\int_a^x b(t) \begin{vmatrix} z_1(x), & z_2(x) \\ z_1(t), & z_2(t) \end{vmatrix}^2 y(t)\, dt\,. \tag{3.26}$$

Using (3.26), we can derive some asymptotic properties of solutions of the differential equation (a).

THEOREM 3.16. Assume that every solution of the second order differential equation (1.4) is bounded in $\langle a, \infty)$ and that the integral $\int_a^x b(t)\, dt$ converges. Then every solution of the differential equation (a) is bounded in $\langle a, \infty)$.

PROOF. Since every solution of (1.4) is bounded in $\langle a, \infty)$, there is a constant $M > 0$ with $|z_1(x)| < M$ and $|z_2(x)| < M$ for $x \geqq a$. In view of the assumption that $\int_a^x |b(t)|\, dt$ converges, $\bar{x} > a$ can be chosen so that $\int_{\bar{x}}^x |b(t)|\, dt < 4^{-1} M^{-4}$. Denoting by $\mu(x)$ the maximum of $|y(t)|$ over $\langle \bar{x}, x \rangle$, by (3.25) we deduce that

$$|y(x)| < M^2(|c_1| + |c_2| + |c_3|) + 2\mu(x) M^4 \int_{\bar{x}}^x |b(t)|\, dt\,,$$

whence, for $x > \bar{x}$, we get

$$|y(x)| < M^2(|c_1| + |c_2| + |c_3|) + \frac{1}{2}\mu(x)\,.$$

Since this equation holds true in the interval $\langle \bar{x}, x \rangle$, it is true also at $x = \xi$, at which $|y(\xi)| = \mu(x)$. Therefore,

$$\mu(x) < M^2(|c_1| + |c_2| + |c_3|) + \frac{1}{2}\mu(x)\,,$$

hence

$$\mu(x) < 2M^2(|c_1| + |c_2| + |c_3|),$$

implying that

$$|y(x)| < 2M^2(|c_1| + |c_2| + |c_3|).$$

Owing to continuity of y in $\langle \alpha, \infty)$, y is bounded there. Thus the theorem is proved. □

THEOREM 3.17. Assume that every solution of the differential equation (1.4) converges to zero as $x \to \infty$ and that $\int_a^x |b(t)| \, dt$ converges. Then every solution of the differential equation (a) converges to zero as $x \to \infty$.

PROOF. Since every solution of (1.4) converges to zero as $x \to \infty$, it is bounded in $\langle \alpha, \infty)$, and by Theorem 3.16, every solution of (a) is bounded. Therefore, there is a constant $M > 0$ with $|y(t)| < M$, $|z_1(t)| < M$, $|z_2(t)| < M$ for $t > \bar{x}$ and $\int_{\bar{x}}^{\infty} b(t) \, dt < M$. Then it follows from (3.26) that

$$|y(x)| < |c_1| z_1^2(x) + |c_2| \, |z_1(x) z_2(x)| + |c_3| z_2^2(x) +$$
$$+ \frac{1}{2} M^4 (z_1^2(x) + 2|z_1(x) z_2(x)| + z_2^2(x)).$$

Since $\lim_{x \to \infty} z_1(x) = \lim_{x \to \infty} z_2(x) = 0$, we have also $\lim_{x \to \infty} y(x) = 0$. Thus the theorem is proved. □

THEOREM 3.18. Let the following assumptions hold in $\langle \alpha, \infty)$.

a) $A(x) > 0$, $\lim_{x \to \infty} A(x) = \infty$.

b) The function $A^{-1/4}(x)$ is convex.

c) The integral $\int_a^x \frac{b(t)}{A(t)} \, dt$ converges.

Then every solution of the differential equation (a) and its first derivative are bounded in $\langle \alpha, \infty)$.

PROOF. Zlámal [146] has proved that under the assumptions a) and b) the following asymptotic formulae are valid for solutions of the second order differential equation (1.4).

$$z(x)=\frac{1}{\sqrt[4]{\frac{1}{2}A(x)}}\left[z_0 \sin\left(\int_\alpha^x \frac{1}{2}A(t)\,\mathrm{d}t+\varphi_0\right)+o(1)\right],$$

$$z'(x)=\sqrt[4]{\frac{1}{2}A(x)}\left[z_0 \cos\left(\int_\alpha^x \frac{1}{2}A(t)\,\mathrm{d}t+\varphi_0\right)+o(1)\right], \tag{3.27}$$

where z_0 and φ_0 are constants.

Choose a constant K so that

$$|z_1(x)|<\frac{K}{\sqrt[4]{\frac{1}{2}A(x)}}$$

and

$$|z_2(x)|<\frac{K}{\sqrt[4]{\frac{1}{2}A(x)}},$$

where z_1, z_2 is a fundamental set for (1.4). In virtue of the assumption c), the lower limit of integration can be chosen to make

$$\int_{\bar{x}}^{\infty} 2\left|\frac{b(t)}{A(t)}\right|\mathrm{d}t<\frac{1}{4K^4}. \tag{3.28}$$

For all $x>\bar{x}$, we conclude from (3.26) and (3.27) that

$$|y(x)|<\frac{K^2}{\sqrt{\frac{1}{2}A(x)}}(|c_1|+|c_2|+|c_3|)+$$

$$+\frac{2K^2}{\sqrt{\frac{1}{2}A(x)}}\int_{\bar{x}}^{x}\frac{K^2}{\sqrt{\frac{1}{2}A(x)}}|b(t)||y(t)|\,\mathrm{d}t. \tag{3.29}$$

Denoting by $M(x)$ the maximum of $|y(t)|\sqrt{\frac{1}{2}A(t)}$ over $\langle \bar{x}, x\rangle$, we have also $|y(x)|\sqrt{\frac{1}{2}A(x)} \leqq M(x)$. The inequality (3.29) may be rewritten in the form

$$|y(x)|\sqrt{\frac{1}{2}A(x)} \leqq M(x) < (|c_1|+|c_2|+|c_3|)K^2 +$$
$$+2K^4 M(x)\int_{\bar{x}}^{x} 2\left|\frac{b(t)}{A(t)}\right| \mathrm{d}t\,.$$

According to (3.28),

$$M(x) \leqq (|c_1|+|c_2|+|c_3|)K^2 + \frac{1}{2}M(x)\,,$$

and finally,

$$|y(x)|\sqrt{\frac{1}{2}A(x)} < 2(|c_1|+|c_2|+|c_3|)K^2\,.$$

Differentiating (3.26), we obtain

$$y'(x) = 2c_1 z_1(x) z_1'(x) + c_2[z_1'(x) z_2(x) + z_1(x) z_2'(x)] +$$
$$+2c_3 z_2(x) z_2'(x) -$$
$$-\int_{\bar{x}}^{x} y(t) b(t) \begin{vmatrix} z_1(x), & z_2(x) \\ z_1(t), & z_2(t) \end{vmatrix} \begin{vmatrix} z_1'(x), & z_2'(x) \\ z_1(t), & z_2(t) \end{vmatrix} \mathrm{d}t\,.$$

Now, from the formulae (3.27) it follows that $|y'(x)|$ is bounded. Thus the theorem is proved. □

In the next three theorems we shall deal with the differential equation (a), written for formal reasons as

$$y''' - py' - \left(\frac{1}{2}p' + b\right) y = 0\,, \qquad (\mathrm{a}_p)$$

where $p=p(x)$, $p'=p'(x)$, $b=b(x)$ are continuous functions of $x\in(a, \infty)$. If $b(x)\equiv 0$ for $x\in(a, \infty)$, we get the self-adjoint differential equation

$$u''' - pu' - \frac{1}{2}p'u = 0 . \tag{3.30}$$

Consider the second order differential equation

$$z'' - \frac{1}{4}pz = 0 . \tag{3.31}$$

We know that if z_1, z_2 is a fundamental set of solutions of (3.31), then the functions $u_1 = z_1^2$, $u_2 = z_1 z_2$, $u_3 = z_2^2$ form a fundamental system for the differential equation (3.30).

Transform the differential equation (a_p) in $\langle \alpha, \infty)$, $a < \alpha < \infty$, by the substitution

$$y = \frac{v(\zeta(x))}{\zeta'(x)} ,$$

where $\zeta(x)$ is a function having a continuous fourth derivative in $\langle \alpha, \infty)$ and $\zeta'(x) > 0$ for $x \in \langle \alpha, \infty)$. The differential equation (a_p) reduces by this transformation to

$$\frac{d^3 v}{d\zeta^3} + P(x)\frac{dv}{d\zeta} + Q(x)v = 0 ,$$

where

$$P(x) = \frac{1}{\zeta'^2(x)}\left[4\sqrt{\zeta'(x)}\left\{\frac{1}{\sqrt{\zeta'(x)}}\right\}'' - p(x)\right] ,$$

$$Q(x) = \frac{1}{2}\frac{dP(x)}{d\zeta} - \frac{b(x)}{\zeta'^2} .$$

Now choose $\zeta(x)$ so that $P(x) \equiv 0$ for $x \in \langle \alpha, \infty)$. It is sufficient to choose $\zeta(x)$ satisfying

$$4\sqrt{\zeta'(x)}\left\{\frac{1}{\sqrt{\zeta'(x)}}\right\}'' - p(x) = 0 ,$$

i.e., if we denote $[\zeta'(x)]^{-1/2} = z(x)$, it is sufficient that $z(x)$ be a solution of the equation (3.31).

The above transformation is admissible in the real domain only if (3.31) has a solution without zeros in (a, ∞). In such a case, (a_p)

reduces to

$$\frac{d^3 v}{d\zeta^3} - \frac{b(x)}{\zeta'^3(x)} v = 0 \tag{3.32}$$

and the transformation takes the form (cf. Sansone [124])

$$y = \frac{v(\zeta(x))}{\zeta'(x)}, \quad \zeta(x) = \int_a^x \frac{1}{z^2(t)} dt . \tag{3.33}$$

If the integral $\int_a^x z^{-2}(t)\, dt$ diverges, then there exist solutions of (3.32) in the whole interval (a, ∞). In the sequel, we shall omit the lower limit of integration for the sake of brevity.

Ghizetti [36] has proved the following result:

Let an n-th order differential equation

$$y^{(n)} = \sum_{r=0}^{n-1} \Phi_r(x) y^{(r)} \tag{3.34}$$

be given and let the integrals

$$\int_a^x |\Phi_r(x)| x^{n-r-1}\, dx ,$$

$r = 0, 1, \ldots, n-1$, converge, where $\Phi_r(x)$ are continuous functions of $x \in (a, \infty)$. Then the asymptotic behaviour of the solutions of (3.34) is the same as that of the solutions of the differential equation $y^{(n)} = 0$, more precisely, there exists a fundamental system of solutions y_k, $k = 1, 2, \ldots, n$, of the differential equation (3.34) such that

$$\lim_{x\to\infty} \frac{y_k^{(s-1)}}{x^{k-s}} = \begin{cases} \dfrac{1}{(k-s)!} & \text{for} \quad s = 1, \ldots, k \\ 0 & \text{for} \quad s > k . \end{cases}$$

Applying this result to the differential equation (3.32), we get, in view of (3.33), the following proposition:

LEMMA 3.7. If the integral

$$\int^\infty \zeta^2(t) \left| \frac{b(t)}{\zeta'^3(t)} \right| d\zeta = \int^\infty \frac{\zeta^2(t)}{\zeta'^2(t)} |b(t)|\, dt$$

converges, then the differential equation (3.32) has a fundamental system of the form

$$v_1 = 1 + o(1)\,,$$
$$v_2 = \zeta[1 + o(1)]\,,$$
$$v_3 = \zeta^2[1 + o(1)]\,,$$

and hence the differential equation (a_p) has a fundamental system

$$\begin{aligned} y_1 &= \frac{1}{\zeta'(x)}[1 + o(1)]\,, \\ y_2 &= \frac{\zeta(x)}{\zeta'(x)}[1 + o(1)]\,, \\ y_3 &= \frac{\zeta^2(x)}{\zeta'(x)}[1 + o(1)]\,. \end{aligned} \tag{3.35}$$

THEOREM 3.19. Let the integrals

$$\int^{\infty} |p(t)|\, dt\,, \tag{3.36}$$

$$\int^{\infty} |b(t)|\, dt \tag{3.37}$$

converge. Then the differential equation

$$y''' - [1 + p(x)]y' - \left[\frac{1}{2}p'(x) + b(x)\right] y = 0 \tag{3.38}$$

has a fundamental system of the form $y_1 = \exp(-x)\ [1 + o(1)]$, $y_2 = 1 + o(1)$, $y_3 = \exp(x)\,[1 + o(1)]$, i.e. the solutions of (3.38) have the same asymptotic behaviour as the solutions of the differential equation

$$y'' - y' = 0\,.$$

PROOF. In our case, the differential equation (3.31) reduces to

$$z'' - \frac{1}{4}[1 + p(x)]\, z = 0\,. \tag{3.39}$$

From (3.36) it follows (Wintner [145]) that a fundamental system for (3.39) has the form $z_1 \sim \exp\left(-\frac{1}{2}x\right)$, $z_2 \sim \exp\left(\frac{1}{2}x\right)$, therefore the equation (3.39) is non-oscillatory and we may put

$$\zeta(x) = \int^x \frac{1}{z_1^2(t)}\,dt = \int^x \frac{dt}{e^{-t/2}[1+o(1)]^2} =$$
$$= \int^x e^t[1+o(1)]\,dt = e^x[1+o(1)]\,.$$

Clearly, $\lim_{x\to\varphi} \zeta(x) = \infty$, $\zeta'(x) > 0$ for sufficiently great x. We show that the hypothesis of Lemma 3.7 is satisfied. In fact,

$$\int^\infty \frac{\zeta^2(t)}{\zeta'^2(t)}\,|b(t)|\,dt = \int^\infty \frac{e^{2t}[1+o(1)]}{e^{2t}[1+o(1)]}\,|b(t)|\,dt =$$
$$= \int^\infty [1+o(1)]\,|b(t)|\,dt \leqq M \int^\infty |b(t)|\,dt$$

by (3.37). Lemma 3.7 implies that the differential equation (3.38) has a fundamental system of solutions having the form

$$y_1 = \frac{1+o(1)}{\zeta'(x)} = \frac{1+o(1)}{e^x[1+o(1)]} = e^{-x}[1+o(1)]\,,$$

$$y_2 = \frac{\zeta(x)}{\zeta'(x)}[1+o(1)] = \frac{e^x[1+o(1)]}{e^x[1+o(1)]}[1+o(1)] = 1+o(1)\,,$$

$$y_3 = \frac{\zeta^2(x)}{\zeta'(x)}[1+o(1)] = \frac{e^{2x}[1+o(1)]}{e^x[1+o(1)]}[1+o(1)] =$$
$$= e^x[1+o(1)]\,.$$

Thus the theorem is proved.□

THEOREM 3.20. Let $\int^\infty |dp(t)| < \infty$, $\lim_{x\to\infty} p(x) = k^2$, where k is a positive constant, and let $\int^\infty |b(t)|\,dt < \infty$.

Then the differential equation (a_p) has a fundamental set of solutions of the form

$$y_1 = e^{-\int^x \sqrt{p(t)}\,dt}[1+o(1)]\,,\quad y_2 = 1+o(1)\,,$$

$$y_3 = e^{\int^x \sqrt{p(t)}\,dt}[1+o(1)]\,.$$

PROOF. It is known (Wintner [145]) that under the assumptions $\int^\infty |dp(t)| < \infty$, $\lim_{x\to\infty} p(x) = k^2$, $k>0$, the differential equation (3.31) has a fundamental set of solutions of the form

$$z_1 \sim e^{-\frac{1}{2}\int^x \sqrt{p(t)}\,dt}, \quad z_2 \sim e^{\frac{1}{2}\int^x \sqrt{p(t)}\,dt}.$$

Put

$$\zeta(x) = \int^x \frac{1}{z_1^2(t)}\,dt = \int^x e^{\int^t \sqrt{p(u)}\,du}[1+o(1)]\,dt =$$

$$= [1+o(1)]\int^x e^{\int^t \sqrt{p(u)}\,du}\,dt\,.$$

Since $\lim_{x\to\infty} p(x) = k^2$, we may put $p(x) = k^2[1+o(1)]$, and thus

$$\zeta(x) = \frac{1+o(1)}{k[1+o(1)]}\int^x \sqrt{p(t)}\,e^{\int^t \sqrt{p(t)}\,du}\,dt =$$

$$= \frac{1}{k}\,[1+o(1)\,e^{\int^x \sqrt{p(t)}\,dt},$$

$$\zeta'(x) = [1+o(1)]\,e^{\int^x \sqrt{p(t)}\,dt}.$$

It is easy to verify that the hypothesis of Lemma 3.7 is satisfied since

$$\int^\infty \frac{\zeta^2(x)}{\zeta'^2(x)}|b(x)|\,dx = \int^\infty \frac{\frac{1}{k^2}[1+o(1)]^2 e^{2\int^x \sqrt{p(t)}\,dt}}{[1+o(1)]^2\,e^{2\int^x \sqrt{p(t)}\,dt}}|b(x)|\,dx =$$

$$= \frac{1}{k^2}\int^\infty [1+o(1)]\,|b(x)|\,dx \leq M\int^\infty |b(x)|\,dx < \infty\,.$$

Therefore, the differential equation (a_p) has a fundamental system of solutions of the form

$$y_1 = \frac{1+o(1)}{\zeta'(x)} = \frac{1+o(1)}{[1+o(1)]\, e^{\int^x \sqrt{p(t)}\, dt}} = e^{-\int^x \sqrt{p(t)}\, dt}\, [1+o(1)] ,$$

$$y_2 = \frac{\zeta(x)}{\zeta'(x)}\, [1+o(1)] = \frac{\frac{1}{k}[1+o(1)]\, e^{\int^x \sqrt{p(t)}\, dt}}{[1+o(1)]\, e^{\int^x \sqrt{p(t)}\, dt}}\, [1+o(1)] =$$

$$= \frac{1}{k}[1+o(1)] ,$$

$$y_3 = \frac{\zeta^2(x)}{\zeta'(x)}\, [1+o(1)] = \frac{\frac{1}{k^2}[1+o(1)]\, e^{2\int^x \sqrt{p(t)}\, dt}}{[1+o(1)]\, e^{\int^x \sqrt{p(t)}\, dt}}\, [1+o(1)] =$$

$$= \frac{1}{k^2}\, e^{\int^x \sqrt{p(t)}\, dt}\, [1+o(1)] .$$

The theorem is thus proved. □

THEOREM 3.21. Let $p(x)>0$, $\int^\infty p(t)\, dt = \infty$, $x \in (a, \infty)$, let

$$\int^\infty \left|\frac{p'(t)}{p(t)}\right|^\nu (\sqrt{p(t)})^{\nu-1}\, dt < \infty, \quad 1 \leqq \nu \leqq 2 ,$$

and let

$$\int^\infty \left|\frac{b(t)}{p(t)}\right| dt < \infty .$$

Then the differential equation (a_p) has a fundamental set of solutions of the form

$$y_1 = \frac{1}{\sqrt{p(x)}}\, e^{-\int^x \sqrt{p(t)}\, dt}\, [1+o(1)] ,$$

$$y_2 = \frac{1}{\sqrt{p(x)}}\, [1+o(1)] ,$$

$$y_3 = \frac{1}{\sqrt{p(x)}} e^{\int^x \sqrt{p(t)}\,dt} [1+o(1)] .$$

PROOF. From the assumptions concerning $p(x)$ it follows that the differential equation (3.31) has a fundamental set of solutions (Hartman and Wintner [63])

$$z_1 \sim \frac{1}{\sqrt[4]{p(x)}} e^{-\frac{1}{2}\int^x \sqrt{p(t)}\,dt}, \quad z_2 \sim \frac{1}{\sqrt[4]{p(x)}} e^{\frac{1}{2}\int^x \sqrt{p(t)}\,dt} .$$

Put

$$\zeta(x) = \int^x \frac{1}{z_1^2(t)}\,dt = \int^x e^{\int^t \sqrt{p(u)}\,du} \sqrt{p(t)}\,[1+o(1)]\,dt =$$
$$= e^{\int^x \sqrt{p(t)}\,dt} [1+o(1)] .$$

Clearly, $\lim_{x\to\infty} \zeta(x) = \infty$ and $\zeta'(x) = e^{\int^x \sqrt{p(t)}\,dt} \sqrt{p(x)}\,[1+o(1)]$. Also, the hypothesis of Lemma 3.7 is fulfilled since

$$\int^\infty \frac{\zeta^2(x)}{\zeta'^2(x)} |b(x)|\,dx =$$
$$= \int^\infty e^{2\int^x \sqrt{p(t)}\,dt} \frac{e^{-2\int^x \sqrt{p(t)}\,dt}}{p(x)} [1+o(1)]\,|b(x)|\,dx =$$
$$= \int^\infty [1+o(1)] \left|\frac{b(x)}{p(x)}\right| dx < M \int^\infty \left|\frac{b(x)}{p(x)}\right| dx .$$

Thus, by Lemma 3.7, the differential equation (a_p) has a fundamental system of solutions of the form

$$y_1 = \frac{1+o(1)}{\zeta'(x)} = \frac{1}{\sqrt{p(x)}} e^{-\int^x \sqrt{p(t)}\,dt} [1+o(1)] ,$$

$$y_2 = \frac{\zeta(x)}{\zeta'(x)} [1+o(1)] =$$
$$= \frac{1}{\sqrt{p(x)}} e^{-\int^x \sqrt{p(t)}\,dt} [1+o(1)]\, e^{\int^x \sqrt{p(t)}\,dt} [1+o(1)] =$$

$$=\frac{1}{\sqrt{p(x)}}[1+o(1)],$$

$$y_3=\frac{\zeta^2(x)}{\zeta'(x)}[1+o(1)]=$$

$$=\frac{1}{\sqrt{p(x)}}e^{-\int^x\sqrt{p(t)}\,dt}\{e^{\int^x\sqrt{p(t)}\,dt}[1+o(1)]\}^2[1+o(1)]=$$

$$=\frac{1}{\sqrt{p(x)}}e^{\int^x\sqrt{p(t)}\,dt}[1+o(1)],$$

so that the theorem is proved. □

§4. BOUNDARY VALUE PROBLEMS

In the four sections of this paragraph, we present a survey of results concerning third order boundary-value problems when the boundary values of a solution of the differential equation (a) are prescribed at two or three points. We begin by introducing the concept of so-called particular Green's functions and their application to solving boundary-value problems (Greguš [50]). Further, we present some results by Sansone [125], who was the first to formulate three-point boundary-value problems. Finally, we develop the so-called Sturm theory for solving third order boundary-value problems (Greguš [41, 45, 47]) and present several special results (Greguš [42, 43, 49, 53—55]).

1. The Green Function and Its Applications

It is well known (Sansone [124]) that if a homogeneous linear differential equation of order n with continuous coefficients in $\langle a, b\rangle$

$$L(y)=0 \tag{4.1}$$

(with the highest-derivative coefficient $a_0(x)\neq 0$) and n boundary conditions

$$U_i(y)=0\,,\quad i=1,2,\ldots,n \tag{4.2}$$

at two points $a<b$ are given and if the problem (4.1), (4.2) has

the trivial solution only, then for every $\xi \in (a, b)$ the so-called Green's function $G(x, \xi)$ can be constructed so that $G(x, \xi)$, $(\partial/\partial x)G(x, \xi)$, ..., $(\partial^{n-2}/\partial x^{n-2})G(x, \xi)$ are continuous functions of $x \in \langle a, b \rangle$ while the function $(\partial^{n-1}/\partial x^{n-1})G(x, \xi)$ is continuous at each $x \in \langle a, b \rangle$ except at $x = \xi$, where it has a discontinuity of the first kind with jump $1/a_0(\xi)$, that is,

$$\frac{\partial^{n-1}}{\partial x^{n-1}} G(\xi+0, \xi) - \frac{\partial^{n-1}}{\partial x^{n-1}} G(\xi-0, \xi) = \frac{1}{a_0(\xi)} . \tag{4.3}$$

Moreover, in the intervals $\langle a, \xi)$ and $(\xi, b\rangle$, the function $G(x, \xi)$ satisfies the differential equation (4.1), together with the boundary conditions (4.2), and is the only one to have these properties.

The function $G(x, \xi)$ has also the property that

$$y(x) = \int_a^b G(x, \xi) r(\xi) \, d\xi \tag{4.4}$$

is a solution of the differential equation $L(y) = r(x)$ ($r(x)$ being a continuous function of $x \in \langle a, b \rangle$) satisfying the boundary conditions (4.2). In Greguš's paper [50], this result is generalized in such a manner that the boundary conditions are prescribed at m points $a_1 < a_2 < \ldots < a_m$ and the coefficients in (4.1) are continuous functions of $x \in \langle a_1, a_m \rangle$. Also, $m-1$ so-called particular Green's functions are constructed corresponding to the intervals $\langle a_i, a_{i+1} \rangle$, $i = 1, \ldots, m-1$, the functions being defined on the whole interval $\langle a_1, a_m \rangle$. Then the formula (4.4) may be written in the form

$$y(x) = \sum_{i=1}^{m-1} \int_{a_i}^{a_{i+1}} G_i(x, \xi) r(\xi) \, d\xi .$$

We shall state this result in a special form for the differential equation (a) and boundary conditions at two or three points, i.e. $m \leqq 3$, and also for special boundary conditions at these points.

Throughout this section we shall assume, for the sake of simplicity, that $A'(x)$, $b(x)$ are continuous functions in $\langle a, \infty)$. The interval is left-closed.

Let $a \leqq b < c < \infty$. Let $\alpha_1, \alpha_2, \alpha_3, \beta_1, \beta_2, \beta_3, \gamma_1, \gamma_2, \gamma_3$ be real numbers with $|\alpha_1| + |\alpha_2| + |\alpha_3| \neq 0$, $|\beta_1| + |\beta_2| + |\beta_3| \neq 0$, $|\gamma_1| + |\gamma_2| + |\gamma_3| \neq 0$. The special boundary conditions, we shall study in the sequel, are as follows:

$$\begin{aligned} U_1(y) &= \alpha_1 y(a) + \alpha_2 y'(a) + \alpha_3 y''(a) = 0 , \\ U_2(y) &= \beta_1 y(b) + \beta_2 y'(b) + \beta_3 y''(b) = 0 , \\ U_3(y) &= \gamma_1 y(c) + \gamma_2 y'(c) + \gamma_3 y''(c) = 0 . \end{aligned} \tag{4.5}$$

If $a = b$, let the first two expressions in (4.5) be linearly independent.

THEOREM 4.1. Let $a < b < c$ and let the problem (a), (4.5) have only the trivial solution. Then for any point $\xi \in \langle a, b\rangle$ or $\xi \in (b, c)$, respectively, there exists a function $G_1(x, \xi)$ or a function $G_2(x, \xi)$, respectively, (particular Green's functions) with the following properties:

a) $G_i(x, \xi)$, $(\partial/\partial x) G_i(x, \xi)$, $i = 1, 2$, are continuous functions of $x \in \langle a, c\rangle$.

b) The function $(\partial^2/\partial x^2) G_i(x, \xi)$, $i = 1, 2$, is a continuous function of $x \in \langle a, c\rangle$ except at the point $x = \xi$, at which it has a discontinuity of the first kind with jump 1, that is,

$$\frac{\partial^2}{\partial x^2} G_i(\xi + 0, \xi) - \frac{\partial^2}{\partial x^2} G_i(\xi - 0, \xi) = 1 . \tag{4.6}$$

c) The functions $G_i(x, \xi)$, $i = 1, 2$, are solutions of the differential equation (a) in $\langle a, \xi)$, $(\xi, c\rangle$ and satisfy the boundary conditions (4.5).

d) The functions $G_i(x, \xi)$, $i = 1, 2$, are uniquely determined by the properties a), b), c).

PROOF. Let y_1, y_2, y_3 be a fundamental system of solutions of the differential equation (a) and let $i = 1$. For $i = 2$, the proof is analogous. In the intervals $\langle a, \xi)$, $(\xi, c\rangle$, the function $G_1(x, \xi)$ necessarily takes the form

$$\begin{aligned} G_1(x, \xi) &= a_1 y_1(x) + a_2 y_2(x) + a_3 y_3(x), \quad a \leqq x < \xi, \\ G_1(x, \xi) &= b_1 y_1(x) + b_2 y_2(x) + b_3 y_3(x), \quad \xi < x \leqq c, \end{aligned}$$

where a_1, a_2, a_3, b_1, b_2, b_3 are suitable constants. From the continuity of $G_1(x, \xi)$ and $(\partial/\partial x) G_1(x, \xi)$ at $x = \xi$ and from the condition (4.6) we obtain the following equations to determine $c_1 = b_1 - a_1$, $c_2 = b_2 - a_2$, $c_3 = b_3 - a_3$:

$$c_1y_1(\xi)+c_2y_2(\xi)+c_3y_3(\xi)=0,$$
$$c_1y_1'(\xi)+c_2y_2'(\xi)+c_3y_3'(\xi)=0\,,$$
$$c_1y_1''(\xi)+c_2y_2''(\xi)+c_3y_3''(\xi)=1\,.$$

The determinant of the above system of equations is the wronskian of a fundamental set of solutions of (a) at ξ, and hence it is non-zero; the numbers c_i, $i=1, 2, 3$, are therefore determined uniquely.

The condition that $G_1(x, \xi)$ is to satisfy the boundary conditions (4.5) is expressed by equations which reduce to

$$b_1U_1(y_1)+b_2U_1(y_2)+b_3U_1(y_3)=$$
$$=c_1U_1(y_1)+c_2U_1(y_2)+c_3U_1(y_3),$$
$$b_1U_2(y_1)+b_2U_2(y_2)+b_3U_2(y_3)\doteq 0,$$
$$b_1U_3(y_1)+b_2U_3(y_2)+b_3U_3(y_3)=0.$$

The determinant of this system of equations is also non-zero because y_1, y_2, y_3 form a fundamental set of solutions of (a), and the problem (a), (4.5) has only the trivial solution. Thus b_1, b_2, b_3 can be determined unambiguously. Since $a_i=b_i-c_i$, $i=1, 2, 3$, the numbers a_i are also uniquely determined. If the right-hand side of the system for calculating b_i were zero, we would have $b_i\equiv 0$, $a_i=-c_i$. Thus the theorem is proved. □

Let the following non-homogeneous differential equation

$$y'''+2A(x)y'+[A'(x)+b(x)]y=r(x) \qquad (a_r)$$

be given, where $r(x)$ is a continuous function of $x\in\langle a, c\rangle$.

THEOREM 4.2. Let $G_1(x, \xi)$, $G_2(x, \xi)$ be Green's functions as specified in Theorem 4.1. Then

$$y(x)=\int_a^b G_1(x, \xi)r(\xi)\,\mathrm{d}\xi+\int_b^c G_2(x, \xi)r(\xi)\,\mathrm{d}\xi \qquad (4.7)$$

is a solution of the differential equation (a_r) satisfying the boundary conditions (4.5).

PROOF. Continuity of the first derivatives of G_1 and G_2 with respect to x implies that

$$y'(x)=\int_a^b \frac{\partial G_1(x,\xi)}{\partial x}\,r(\xi)\,d\xi+\int_b^c \frac{\partial G_2(x,\xi)}{\partial x}\,r(\xi)\,d\xi\,.$$

We now calculate the derivatives $y''(x)$ and $y'''(x)$, considering separately the cases, when x is interior to one of the intervals (a, b), (b, c), and when it lies at an endpoint of one of these intervals. To be specific, let $x\in(a, b)$. Then

$$y''(x)=\frac{d}{dx}\Big[\int_a^x \frac{\partial}{\partial x}\,G_1(x,\xi)r(\xi)\,d\xi+$$

$$+\int_x^b \frac{\partial}{\partial x}\,G_1(x,\xi)r(\xi)\,d\xi+\int_b^c \frac{\partial^2}{\partial x^2}\,G_2(x,\xi)r(\xi)\,d\xi\Big]=$$

$$=\int_a^x \frac{\partial^2}{\partial x^2}\,G_1(x,\xi)r(\xi)\,d\xi+\frac{\partial}{\partial x}\,G_1(x,x)r(x)+$$

$$+\int_x^b \frac{\partial^2}{\partial x^2}\,G_1(x,\xi)r(\xi)\,d\xi-\frac{\partial}{\partial x}\,G_1(x,x)r(x)+$$

$$+\int_b^c \frac{\partial^2}{\partial x^2}\,G_2(x,\xi)r(\xi)\,d\xi=$$

$$=\int_a^b \frac{\partial^2}{\partial x^2}\,G_1(x,\xi)r(\xi)\,d\xi+\int_b^c \frac{\partial^2}{\partial x^2}\,G_2(x,\xi)r(\xi)\,d\xi\,.$$

Similarly,

$$y'''(x)=\int_a^b \frac{\partial^3}{\partial x^3}\,G_1(x,\xi)r(\xi)\,d\xi+r(x)\Big[\frac{\partial^2}{\partial x^2}\,G_1(x,x-0)-$$

$$-\frac{\partial^2}{\partial x^2}\,G_1(x,x+0)\Big]+\int_b^c \frac{\partial^3}{\partial x^3}\,G_2(x,\xi)r(\xi)\,d\xi=$$

$$=\int_a^b \frac{\partial^3}{\partial x^3}\,G_1(x,\xi)r(\xi)\,d\xi+r(x)+$$

$$+\int_b^c \frac{\partial^3}{\partial x^3}\,G_2(x,\xi)r(\xi)\,d\xi,$$

where we have made use of (4.6) and the fact that

$$\frac{\partial^2}{\partial x^2}\,G_1(x,x-0)-\frac{\partial^2}{\partial x^2}\,G_1(x,x+0)=$$

$$=\frac{\partial^2}{\partial x^2}G_1(x+0,x)-\frac{\partial^2}{\partial x^2}G_1(x-0,x)=1\,.$$

The derivatives of $y(x)$ when $x\in(b,c)$ are found analogously.

From the expressions for $y''(x)$ and $y'''(x)$ we immediately deduce that the limits $\lim_{x\to a^+} y(x)$ and $\lim_{x\to a^+} y'''(x)$, hence also $\lim_{x\to b^-} y''(x)$, $\lim_{x\to b^-} y'''(x)$, $\lim_{x\to b^+} y''(x)$, $\lim_{x\to b^+} y'''(x)$, $\lim_{x\to c^-} y(x)$, $\lim_{x\to c^-} y'''(x)$ all exist and $\lim_{x\to b^+} y''(x)=\lim_{x\to b^-} y''(x)=y''(b)$.

We now substitute in the differential equation (a_r) and the boundary conditions (4.5) the expressions for y', y'', y''' and y we have found; it is easy to verify that these become identities, so the theorem is proved. □

REMARK 4.1. If $b=a$, then the assertions of Theorems 4.1, 4.2 hold true, but there exists exactly one Green's function $G_1(x,\xi)=G(x,\xi)$. With this exception, the proofs of Theorems 4.1 and 4.2 are the same.

COROLLARY 4.1. Let the homogeneous problem (a), (4.5) be unsolvable, i.e. let it have only the trivial solution $y\equiv 0$. Let $G_i(x,\xi)$, $i=1,2$, be the Green functions for the problem (a), (4.5). Also, let u_1, u_2 and u_3 be solutions of (a) satisfying the boundary conditions

$$U_1(u_1)=1,\quad U_2(u_1)=0,\quad U_3(u_1)=0,$$
$$U_1(u_2)=0,\quad U_2(u_2)=1,\quad U_3(u_2)=0,$$
$$U_1(u_3)=0,\quad U_2(u_3)=0,\quad U_3(u_3)=1.$$

(Evidently, such solutions do exist (Sansone [124]).)

Then the function

$$y(x)=\int_a^b G_1(x,\xi)r(\xi)\,d\xi+\int_b^c G_2(x,\xi)r(\xi)\,d\xi+$$
$$+\gamma_1u_1(x)+\gamma_2u_2(x)+\gamma_3u_3(x)$$

satisfies the differential equation (a_r) and the conditions

$$U_i(y)=\gamma_i,\quad i=1,2,3.$$

This assertion can be proved as in Theorem 4.2.

COROLLARY 4.2. Let a third order differential equation be given in the form

$$y''' + 2A(x)y' + [A'(x) + b(x)]y = \lambda g(x)y\,, \qquad (a_\lambda)$$

where λ is a parameter and $g(x)$ is a continuous function on $\langle a, c\rangle$. Moreover, let the problem (a), (4.5) have only the trivial solution and let $G_1(x, \xi)$, $G_2(x, \xi)$ be its Green functions. The question of existence of a solution of the problem (a_λ), (4.5) is equivalent to the question of existence of a solution of the integral equation

$$y(x) = \lambda\left[\int_a^b G_1(x, \xi)g(\xi)y(\xi)\,d\xi + \right.$$

$$\left. + \int_b^c G_2(x, \xi)g(\xi)y(\xi)\,d\xi\right].$$

The assertion now follows from Theorem 4.2.

The following may serve as a simple example. Solve the problem

$$v''' = g(x), \quad v(0) = v'(0) = v(c) = 0\,, \qquad (4.8)$$

assuming that $a = 0$, $b = 0$ in the above notation.

The only solution of (4.8) is

$$v(x) = \int_0^c G(x, \xi)g(\xi)\,d\xi\,,$$

where $G(x, \xi)$ is the Green function of the homogeneous problem corresponding to (4.8), which can be found as follows.

x^2, x, 1 is a fundamental system of solutions of the differential equation $v''' = 0$ with wronskian $W = -2$. We require

$$G(x, \xi) = \begin{cases} a_1x^2 + a_2x + a_3, & 0 \leqq x \leqq \xi, \\ b_1x^2 + b_2x + b_3, & 0 < \xi \leqq x \leqq c. \end{cases}$$

At ξ we have

$$c_1\xi^2 + c_2\xi + c_3 = 0\,,$$

$$2c_1\xi + c_2 = 0\,,$$

$$2c_1 = 1, \qquad c_i = b_i - a_i, \; i = 1, 2, 3.$$

Consequently, $c_1 = \frac{1}{2}$, $c_2 = -\xi$, $c_3 = \frac{1}{2}\xi^2$.

From the requirement that $G(x, \xi)$ should satisfy the boundary conditions, it follows that

$$a_3=0\,,$$
$$a_2=0\,,$$
$$b_1c^2+b_2c+b_3=0\,,$$

hence

$$a_1=\frac{\xi}{c}-\frac{\xi^2}{2c}-\frac{1}{2}\,,\quad b_1=-\frac{\xi^2}{2c^2}+\frac{\xi}{c}\,,\quad b_2=-\xi,\quad b_3=\frac{1}{2}\xi^2.$$

Thus we obtain the following form of the Green function:

$$G(x,\xi)=\begin{cases}-\dfrac{1}{2}\left(1-\dfrac{\xi}{c}\right)^2 x^2, & 0\leqq x\leqq \xi\leqq c,\\[2mm] \dfrac{\xi}{c}\left(1-\dfrac{\xi}{2c}\right)x^2-\xi x+\dfrac{1}{2}\xi^2, & 0<\xi\leqq x\leqq c\,.\end{cases} \tag{4.9}$$

Similarly we could derive the corresponding integral equation for solving the problem

$$v'''=\lambda g(x)v,\quad v(0)=v(c)=0,\quad v'(0)=0,$$

which is of the form

$$v(x)=\lambda\int_a^b G(x,\xi)g(\xi)v(\xi)\,d\xi\,,$$

where $G(x, \xi)$ is again given by (4.9).

2. *Further Applications of Integral Equations to the Solution of Boundary-value Problems*

Consider the third order differential equation

$$y'''+2\lambda A(x)y'+\lambda[A'(x)+b(x)]y=0 \tag{a_{λ_1}}$$

and the boundary conditions

$$y(a)=y'(a)=0\,,\quad y(b)=0\,,\quad a<b<\infty\,, \tag{4.10}$$

where $-\infty<\lambda<\infty$ is a parameter.

Integrate the differential equation (a_{λ_1}) from a to $x \leqq b$. We get

$$y''(x)+\lambda[2A(x)y(x)+$$
$$+\int_a^x [b(t)-A'(t)]y(t)\,dt]-y''(a)=0\,. \tag{4.11}$$

It is easy to verify that the Green function $G(x, \xi)$ for the boundary problem

$$y''=0\,, \quad y(a)=y(b)=0$$

is given by

$$G(x, \xi)=\begin{cases} \dfrac{(b-\xi)(x-a)}{b-a}\,, & a\leqq x\leqq \zeta\,, \\ \dfrac{(\xi-a)(b-x)}{b-a}\,, & \xi\leqq x\leqq b\,, \end{cases} \tag{4.12}$$

where $a<\xi<b$.

According to (4.4) of the preceding paragraph, a solution of (4.11) satisfying $y(a)=y(b)=0$ simultaneously satisfies the integral equation

$$y(x)=\lambda\int_a^b G(x, \xi)\,[2A(\xi)y(\xi)+$$
$$+\int_a^\xi [b(t)-A'(t)]y(t)\,dt]\,d\xi-y''(a)\int_a^b G(x, \xi)\,d\xi\,. \tag{4.13}$$

It follows from (4.12) that

$$\int_a^b G(x, \xi)\,d\xi=\int_a^\xi G(x, \xi)\,d\xi+\int_\xi^b G(x, \xi)\,d\xi=$$
$$=\frac{b-x)(x-a)}{2}\,.$$

Denote

$$\int_t^b G(x, \xi)\,d\xi=H(x, t)\,. \tag{4.14}$$

Then

$$\int_a^b G(x, \xi)\,d\xi\int_a^\xi [b(t)-A'(t)]\,y(t)\,dt=$$

$$=\int_a^b H(x, t)\,[b(t)-A'(t)]\,y(t)\,dt$$

and from (4.13) we obtain

$$y(x)=\lambda\int_a^b [2G(x,t)A(t)+H(x,t)(b(t)-A'(t))]y(t)\,dt-$$
$$-y''(a)\frac{(b-x)\,(x-a)}{2}. \tag{4.15}$$

By (4.14) and (4.12) we obtain

$$H(x, t)=\int_t^x G(x, \xi)\,d\xi+\int_x^b G(x, \xi)\,d\xi=$$
$$=\frac{(b-x)\,(x-a)}{2}-\frac{(b-x)\,(t-a)^2}{2(b-a)}, \quad x\geqq t, \tag{4.16}$$

$$H(x, t)=\frac{(x-a)\,(b-t)^2}{2(b-a)}, \quad x<t. \tag{4.17}$$

From (4.15), after decomposing the integral on the right-hand side into two integrals over (a, x) and (x, b), respectively, and taking (4.12), (4.16) and (4.17) into account, it is easy to calculate the derivative $y'(x)$. The condition $y(a)=0$ implies that

$$0=-y''(a)\frac{b-a}{2}+2\lambda\int_a^b \frac{b-t}{b-a}A(t)y(t)\,dt+$$
$$+\lambda\int_a^b \frac{(b-t)^2}{2(b-a)}[b(t)-A'(t)]\,y(t)\,dt.$$

Thus we have obtained a formula for $y''(a)$. Substituting this $y''(a)$ in (4.15) yields the final result, that is, we have proved the following theorem.

THEOREM 4.3. A solution y of the differential equation (a_{λ_1}) satisfying the boundary conditions (4.10) is, at the same time, a solution of Fredholm's integral equation of the second kind (Mikhlin [97]).

$$y(x)=\int_a^b K(x, t)y(t)\,dt,$$

where

$$K(x,t)=\frac{b-x}{2(b-a)}[\{(x-a)(b-t)+(b-a)(x-t)\}\{b(t)-$$

$$-A'(t)\}(t-a)+\{-b(x-t)+t(x-a)-$$

$$-a(t-a)\}\,4a(t)]\,,\qquad t\leqq x\,,$$

$$K(x,t)=\frac{2(b-t)(x-a)^2}{(b-a)^2}A(t)+\frac{(x-a)^2(b-t)^2}{2(b-a)^2}\cdot$$

$$\cdot[b(t)-A'(t)]\,m,\qquad t\geqq x\,.$$

REMARK 4.2. By employing the theory of integral equations, Sansone [125] has derived from Theorem 4.3 conditions under which the boundary-value problem (a_{λ_1}), (4.10) has one eigenvalue, infinitely many eigenvalues, or no eigenvalues.

3. Generalized Sturm Theory for Third Order Boundary-value Problems

We shall study the third order differential equation

$$y'''+2A(x,\lambda)y'+[A'(x,\lambda)+b(x,\lambda)]\,y=0\,, \qquad (a_\Delta)$$

where $A(x,\lambda)$, $(\partial/\partial x)A(x,\lambda)=A'(x,\lambda)$, $b(x,\lambda)$ are continuous functions of $x\in\langle a,\infty)$ and $\lambda\in(\Lambda_1,\Lambda_2)$.

The following boundary conditions will be prescribed at three points, $a\leqq b<c$,

$$\begin{aligned}&y^{(i)}(a,\lambda)=0\,,\quad i=0,1,2\\&\alpha_1 y(b,\lambda)-\alpha y'(b,\lambda)=0\,,\\&\beta_1 y(c,\lambda)-\beta y'(c,\lambda)=0\,,\end{aligned} \qquad (4.18)$$

where α, α_1, β, β_1 are either constants or continuous functions of the parameter $\lambda\in(\Lambda_1,\Lambda_2)$.

An oscillation theorem analogous to the Sturm oscillation theorem for second order differential equations (Sansone [124]) was first established by Sansone [125]. The theorem, adapted for the differential equation (a_Δ), reads as follows.

THEOREM A. Let $A(x, \lambda)>0$, $b(x, \lambda)$ have the property (v) in $\langle a, \infty)$ for all $\lambda \in (\Lambda_1, \Lambda_2)$, and let $\lim_{\lambda \to \Lambda_2} A(x, \lambda)=\infty$ uniformly for all $x \in \langle a, \infty)$. Let $a<b<\infty$ and let $y(x, \lambda)$ be a solution of the differential equation (a_Δ) with

$$y(a, \lambda)y''(a, \lambda)-\frac{1}{2}y'^2(a, \lambda)+A(a, \lambda)y^2(a, \lambda)\leqq 0$$

for $\lambda \in (\Lambda_1, \Lambda_2)$. As $\lambda \to \Lambda_2$, the number of zeros of $y(x, \lambda)$ in (a, b) increases to infinity and, moreover, the distances between any two neighbouring zeros tend to zero.

The proof of this theorem follows from comparison of the differential equation (a_Δ) with the self-adjoint equation

$$y'''+2A(x, \lambda)y'+A'(x, \lambda)y=0 .$$

Another oscillation theorem was proved by Greguš [45] and reformulated as follows (Greguš [47]).

THEOREM B. Let $|A(x, \lambda)|\leqq k$, $|A'(x, \lambda)|\leqq k$, $k>0$, for $x \in (a, \infty)$ and $\lambda \in (\Lambda_1, \Lambda_2)$ and let $\lim_{\lambda \to \Lambda_2} b(x, \lambda)=\infty$ uniformly for all $x \in \langle a, \infty)$. Let $a<b$ and let $y(x, \lambda)$ be a non-trivial solution of the differential equation (a_Δ) with $y(a, \lambda)=0$. As $\lambda \to \Lambda_2$, the number of zeros of $y(x, \lambda)$ in (a, b) increases to infinity, while the differences between any two neighbouring zeros of the solution y tend to zero.

This theorem has been proved using the properties of bands and comparison of the differential equation (a_Δ) with the differential equation

$$z'''+2A(x, \lambda)z'+[A'(x, \lambda)+n^3+2A(x, \lambda)n+ \\ +A'(x, \lambda)]\, z=0 ,$$

which is obtained by differentiating the second order equation

$$e^{nx}z''-n\, e^{nx}z'+e^{nx}(n^2+2A)z=0 ,$$

where n is a positive integer and the last second order equation, being the equation of the band at $-\infty$, is oscillatory for sufficiently large n.

We shall not describe the proofs of these theorems in any more detail, because we shall present here an oscillation theorem (Rovder [122, 123]) which generalizes in a sense both Theorem A and Theorem B.

For this purpose, we prove some auxiliary assertions.

DEFINITION. A solution y of the differential equation (a) is said to belong to class $d(k)$ in $\langle a, \infty)$ if the difference between any two consecutive zeros of y in $\langle a, \infty)$ is less than a positive real number k.

THEOREM 4.4. Let $b(x) \geqq 0$ for $x \in \langle a, \infty)$ and let there exist at least one oscillatory solution of the differential equation (a) belonging to class $d(k)$ in $\langle \alpha, \infty)$, $a \leqq \alpha$. Then, for some $K > 0$, every solution of (a) having a zero in $\langle \alpha, \infty)$ belongs to class $d(K)$ in $\langle \alpha, \infty)$.

PROOF. Let y be a solution of (a), oscillatory in $\langle \alpha, \infty)$ and belonging to class $d(k)$ in $\langle \alpha, \infty)$. Let α be a simple zero of the solution y. Let z be a solution of (a) with a double zero at α. Since Theorem 1.14 implies that every solution having a zero oscillates in $\langle \alpha, \infty)$, z is oscillatory whenever some solution is oscillatory. Let $\alpha \leqq x_1 < x_2 < x_3$ be consecutive zeros of the solution y. Then z must have a zero in $(x_1, x_3\rangle$, because if z were (say) positive in $(x_1, x_3\rangle$, it would be positive in $\langle x_2, x_3\rangle$ as well. Then by Lemma 1.2 there would exist a constant c and some $\tau \in (x_2, x_3)$ such that the solution $w(x) = z(x) - cy(x)$ of (a) would have a double zero at τ, contradicting the integral identity (1.6) for the solution w. In fact,

$$\left[ww'' - \frac{1}{2} w'^2 + Aw^2\right]_\alpha^\tau = -\int_\alpha^\tau b(t) w^2(t)\, dt\,.$$

Therefore, z has a zero in (x_1, x_3). Since $x_1 < x_2 < x_3$ are any three consecutive zeros of y, the solution z is in class $d(3k)$ in $\langle \alpha, \infty)$.

Now let $u(x)$ be a solution of (a) having a simple zero at α. Theorem 2.57 implies that the zeros of u and z separate each other in (α, ∞) in the sense that if $\xi_1 < \xi_2$ are two neighbouring zeros of z, then u has at least one zero between ξ_1 and ξ_2. Thus the solution u is in class $d(6k)$ in $\langle \alpha, \infty)$. If the solution y had a double zero at α, we would conclude (analogously to the preceding case) that every solution having a zero at α belongs to the class $d(6k)$ in $\langle\alpha, \infty)$.

Assume now that v is a solution of the differential equation (a) vanishing at β and that β is different from the zeros of y. Then, for some solution $w(x)$ of the differential equation (a), $w(\alpha)=w(\beta)=0$. Since w and y have a common zero at α, the solution w is in class $d(6k)$ in $\langle\alpha, \infty)$. Both v and w have a zero at β, v belongs to class $d(36k)$ in $\langle\alpha, \infty)$. If we put $36k=K$, the theorem is proved. □

LEMMA 4.1. Let $p, q\geqq 0$ be real numbers and let the differential equation

$$z'''+pz'+\frac{q}{2}z=0 \tag{4.19}$$

be oscillatory in $\langle\alpha, \infty)$. Then every oscillatory solution of (4.19) is in class $d(K/r)$ in $\langle a, \infty)$, $a\leqq\alpha$, where K is a number independent of p, q and

$$r=(-q+h)^{1/3}+(q+h)^{1/3},\quad h=\left(q^2+\frac{16}{27}p^3\right)^{1/2}. \tag{4.20}$$

PROOF. If the differential equation (4.19) is oscillatory in $\langle\alpha, \infty)$, its characteristic equation has the following roots:

$$x_1=u+v,\quad x_{2,3}=r_1\pm r_2\frac{\sqrt{3}}{2}i,\quad i=\sqrt{-1},$$

where

$$u=\left\{-\frac{1}{4}q+\left[\left(\frac{1}{4}q\right)^2+\left(\frac{1}{3}p\right)^3\right]^{1/2}\right\}^{1/3},$$

$$v=\left\{-\frac{1}{4}q-\left[\left(\frac{1}{4}q\right)^2+\left(\frac{1}{3}p\right)^3\right]^{1/2}\right\}^{1/3}$$

and $r_1=-2^{-1}(u+v)$, $r_2=u-v$. Since $(4^{-1}q)^2+(3^{-1}p)^3>0$, the numbers u, v, r_1 and r_2 are real. Let $h=[q^2+(16/27)p^3]^{1/2}$. Then r_2 may be written in the form

$$r_2=\frac{1}{\sqrt[3]{4}}[(-q+h)^{1/3}+(q+h)^{1/3}].$$

Denoting the expression in brackets by r, for the roots $x_{2,3}$ we get

$$x_{2,3}=r_1\pm\frac{\sqrt[3]{6}}{4}\,ri\,.$$

One of the solutions of (4.19) has the form

$$y(x)=c_1\,\mathrm{e}^{r_1x}\sin\frac{\sqrt[3]{6}}{4}\,r(x-\alpha)\,.$$

It follows that for every $\alpha\geqq a$ there is a solution of (4.19) belonging to class $d\left(\frac{4\pi}{\sqrt[3]{6}}\frac{1}{r}\right)$ in $\langle\alpha,\infty)$. By Theorem 4.4, every solution having a zero is in class $d(Kr)$ in $\langle\alpha,\infty)$, where $K=144\pi 6^{-1/3}$, and the lemma is proved. □

THEOREM 4.5 (Oscillation Theorem). Let the coefficients of differential equation (a_Δ) satisfy the following conditions:

a) $A(x,\lambda)\geqq p/2$ for all $x\in\langle a,\infty)$ and $\lambda\in(\Lambda_1,\Lambda_2)$, where p is a real constant, and moreover, $\lim\limits_{\lambda\to\Lambda_2} b(x,\lambda)=\infty$ uniformly for all $x\in\langle a,\infty)$,

or

b) $b(x,\lambda)\geqq 0$ for all $x\in\langle a,\infty)$ and $\lambda\in(\Lambda_1,\Lambda_2)$, and moreover, $\lim\limits_{\lambda\to\Lambda_2} A(x,\lambda)=\infty$ uniformly for all $x\in\langle a,\infty)$.

Let $a<b<\infty$. Let $y(x,\lambda)$ be a non-trivial solution of the differential equation (a_Δ) with $y(a,\lambda)=0$. With increasing $\lambda\to\Lambda_2$, the number of zeros of y in $\langle a,b\rangle$ increases to infinity and, at the same time, the distance between any two neighbouring zeros of y converges to zero.

PROOF. Suppose the conditions a) hold true. For every $q>0$, there is $\lambda_0\in(\Lambda_1,\Lambda_2)$ such that for $\lambda>\lambda_0\in(\Lambda_1,\Lambda_2)$ we have $b(x,\lambda)>q/2$ for all $x\in\langle a,\infty)$.

Let q be such that the differential equation (4.19) is oscillatory in $\langle a,\infty)$. A solution z of (4.19) with $z(a)=z'(a)=0$, $z''(a)\neq 0$ is also oscillatory by Theorem 1.14. According to Lemma 4.1, the solution z is in class $d(K/r)$ in $\langle a, \infty)$, where r is given by (4.20). If $b(x, \lambda)$ diverges uniformly to ∞ in $\langle a, \infty)$ and, at the same time $q \to \infty$ while $b(x, \lambda) > q/2$ for all $x \in \langle a, \infty)$, then also $r \to \infty$ by (4.20). The number of zeros of z in $\langle a, b \rangle$ increases to infinity and the distances between any two zeros of z converge to zero.

Let $y(x, \lambda)$ be a solution of (a_Δ) having a simple zero at a. Let $a < x_1$ be two consecutive zeros of a solution z of (4.19). Since for all $x \in \langle a, \infty)$ and $\lambda \in (\Lambda_0, \Lambda_2)$ we have

$$A(x, \lambda) \geqq \frac{p}{2}, \quad b(x, \lambda) \geqq \frac{q}{2} > 0,$$

according to the Comparison Theorem 2.57, the solution y has at least one zero a_1 in $(a, x_1\rangle$ and evidently $|a - a_1| \leqq |x_1 - a| < K/r$.

Now let $z_1(x)$ be a solution of (4.19) with $z_1(a_1) = z_1'(a_1) = 0$, $z_1''(a_1) \neq 0$ and let $x_2 > a_1$ be its first zero to the right of a_1. By Theorem 2.57, the solution $y(x, \lambda)$ vanishes again at some $a_2 \in (a_1, x_2\rangle$ and $|a_1 - a_2| \leqq |a_1 - x_2| < K/r$. Using mathematical induction we show that the distance between any two neighbouring zeros of $y(x, \lambda)$ is less than K/r.

So far we have assumed that a is a simple zero of the solution $y(x, \lambda)$ of the differential equation (a_Δ). In view of the assumption that $b(x, \lambda) \geqq 0$ for all x and λ, it follows that all the zeros of $y(x, \lambda)$ are simple. If $y(x, \lambda)$ has a double zero at a, that is to say $y(a, \lambda) = y'(a, \lambda) = 0$, $y''(a, \lambda) \neq 0$, then by Lemma 4.1 the distance between any two neighbouring zeros of y is less than K_1/r, where $K_1 = 6K$. Therefore the distance between any two neighbouring zeros of any non-trivial solution $y(x, \lambda)$ of the differential equation (a_Δ) with $y(a, \lambda) = 0$ is less than K_1/r. From this fact and that the assumptions

$$\lim_{\lambda \to \Lambda_2} b(x, \lambda) = \infty \quad \text{uniformly for } x \in \langle a, \infty)$$

and $q \to \infty$ while $b(x, \lambda) > q/2$ it follows that $r \to \infty$, and so with $\lambda \to \infty$ the number of zeros of $y(x, \lambda)$ in $(a, b\rangle$ increases to infinity while the distance between any two neighbouring zeros of y tends to zero. Thus the assertion of the theorem under the assumption a) is proved.

Let the assumption b) hold true. Then, given any $p>0$, there exists such a λ_0 that for $\lambda>\lambda_0\in(\Lambda_1, \Lambda_2)$ we have $A(x, \lambda)>p/2$ for all $x\in\langle a, \infty)$.

The differential equation

$$z'''+pz'=0 \tag{4.21}$$

is oscillatory in $\langle a, \infty)$ and, as $p\to\infty$, the distance of any two neighbouring zeros of a solution z of (4.21) with $z(a)=z'(a)=0$, $z''(a)\neq 0$ tends to zero.

As in the case a), it can be proved that every non-trivial solution $y(x, \lambda)$ of the differential equation (a_Δ) with $y(a, \lambda)=0$ has also the property that the number of its zeros in $\langle a, b)$ increases to infinity as $\lambda\to\Lambda_2$ and the distances between any two neighbouring zeros of $y(x, \lambda)$ tend to zero. Thus the theorem is proved. □

COROLLARY 4.3. Let the hypotheses of Theorem 4.5 be satisfied and, moreover, let $b(x, \lambda)$ have the property (v) for $x\in\langle a, \infty)$ and $\lambda\in(\Lambda_1, \Lambda_2)$. Then the assertion of Theorem 4.5 holds true; moreover, the zeros of any two independent solutions of the differential equation (a_Δ) with $y(a, \lambda)=0$ separate each other.

The proof follows from the fact that the solutions $y(x, \lambda)$ with $y(a, \lambda)=0$ form a band at a and that the differential equation (a_Δ) is in class I. □

Theorem 4.5 represents a particular case of the following more general theorem.

THEOREM 4.6. Let the function $A(x, \lambda)$ be bounded below in $\langle a, \infty)$ for all $\lambda\in(\Lambda_1, \Lambda_2)$. Let $b(x, \lambda)\geqq 0$ for $x\in(\Lambda_1, \Lambda_2)$. Further, let

$$\frac{p(\lambda)}{2}=\inf_{a\leq x<\infty} A(x, \lambda), \quad \frac{q(\lambda)}{2}=\inf_{a\leq x<\infty} b(x, \lambda),$$

$$\lambda\in(\Lambda_1, \Lambda_2),$$

$$h(\lambda)=q^2(\lambda)+\frac{16}{27}p^3(\lambda), \tag{4.22}$$

$$r(\lambda)=[-q(\lambda)+h(\lambda)]^{1/3}+[q(\lambda)+h(\lambda)]^{1/3} \tag{4.23}$$

and let $\lim\limits_{\lambda\to\Lambda_2} r(\lambda)=\infty$.

Then the assertion of Theorem 4.5 holds true.

PROOF. The assumption $\lim_{\lambda \to \Lambda_2} r(\lambda) = \infty$ implies that there exists $\lambda_0 \in (\Lambda_1, \Lambda_2)$ such that $r(\lambda) < 0$ for $\lambda > \lambda_0 \in (\Lambda_1, \Lambda_2)$. It follows that $h(\lambda) > 0$ for $\lambda \in (\lambda_0, \Lambda_2)$, because (4.23) for $h(\lambda) \leqq 0$ implies $r(\lambda) \leqq 0$. The differential equation

$$z''' + p(\lambda) z' + \frac{q(\lambda)}{2} z = 0 \tag{4.24}$$

is therefore oscillatory for every $\lambda \in (\lambda_0, \Lambda_2)$. Since $\lim_{\lambda \to \Lambda_2} r(\lambda) = \infty$, the distance between any two neighbouring zeros of $z(x, \lambda)$ having a double zero at a tends to zero as $\lambda \to \Lambda_2$. By the definitions of $p(\lambda)$ and $q(\lambda)$, we have $A(x, \lambda) \geqq p(\lambda)/2$, $b(x, \lambda) \geqq q(\lambda)/2 \geqq 0$. Thus the hypotheses of Theorem 2.57 are fulfilled. The proof now continues in the same way as that of Theorem 4.5. □

LEMMA 4.2. Let $b(x, \lambda)$ have the property (v) for $x \in \langle a, \infty)$ and $\lambda \in (\Lambda_1, \Lambda_2)$. Let $y(x, \lambda)$ be a non-trivial solution of the differential equation (a_Δ) with $y(a, \lambda) = 0$. Then the zeros of $y(x, \lambda)$ in (a, ∞) (if they exist) are continuous functions of the parameter $\lambda \in (\Lambda_1, \Lambda_2)$.

PROOF. From a general theorem on first order differential systems whose right-hand sides are functions of a parameter λ (Sansone [124]) it follows for every solution y of (a_Δ) that, in any closed two-dimensional domain of x, λ which is a subset of the interval $\langle a, \infty) \times (\Lambda_1, \Lambda_2)$, the functions $y(x, \lambda)$, $y'(x, \lambda)$ and $y''(x, \lambda)$ are continuous in x and λ.

Let $y(x, \lambda)$ be a non-trivial solution of (a_Δ) with $y(a, \lambda) = 0$ and let all its zeros in $\langle a, \infty)$ be

$$a < x_1(\lambda) < x_2(\lambda) < \ldots < x_\nu(\lambda) < \ldots .$$

It is clear that $y(x_i(\lambda_1), \lambda_1) = 0$ but $y'(x_i(\lambda_1), \lambda_1) \neq 0$, $\Lambda_1 < \lambda_1 < \Lambda_2$.

Then for any sufficiently small $\varepsilon > 0$ there exists $\delta > 0$ such that for

$|\lambda-\lambda_1|<\delta$ and $a \leqq x \leqq x_\nu(\lambda_1)+\varepsilon$ we have $|y(x, \lambda_1)-y(x, \lambda)|< \min\limits_u |y(u, \lambda_1)|$, where

$$a+\varepsilon \leqq u \leqq x_1(\lambda_1)-\varepsilon\,,$$
$$x_1(\lambda_1)+\varepsilon \leqq u \leqq x_2(\lambda_1)-\varepsilon\,,$$
$$\cdots\cdots\cdots\cdots\cdots\cdots\cdots\cdots\cdots\cdots\cdots\cdots$$
$$x_{\nu-1}(\lambda_1)+\varepsilon \leqq u \leqq x_\nu(\lambda_1)-\varepsilon\,,$$
$$x_\nu(\lambda_1)+\varepsilon = u\,.$$

Then, evidently, the function $y(x, \lambda)$ for $|\lambda-\lambda_1|<\delta$ has exactly one zero in every interval $[x_i(\lambda_1)-\varepsilon, x_i(\lambda_1)+\varepsilon], i=1,2,\ldots,$ and no zeros in the intervals $(a, x_1(\lambda_1)-\varepsilon)$, $(x_i(\lambda_1)+\varepsilon,\ x_{i+1}(\lambda_1)-\varepsilon)$, $i=1, 2, \ldots, \nu$.

This implies that the zeros $x_\nu(\lambda)$ depend continuously on the parameter λ, so that the theorem is proved. □

REMARK 4.3. It can be proved similarly that the zeros of the derivative $y'(x, \lambda)$ with respect to x as well as those of the second derivative $y''(x, \lambda)$ are continuous functions of the parameter $\lambda \in (\Lambda_1, \Lambda_2)$.

REMARK 4.4. Precisely as in Lemma 4.2, it can be proved that the zeros of a non-trivial solution $y(x, \lambda)$ of the differential equation (a_Δ) with $y'(a, \lambda)=0$ or with $y''(a, \lambda)=0$ are continuous functions of the parameter λ provided that the band of the second kind or, respectively, the band of the third kind at a is regular in (a, ∞) (cf. §1—6), i.e. if the respective functions $w_2(x, \lambda)$ or $w_3(x, \lambda)$ in the differential equation (c), regarded as solutions of the differential equation adjoint to (a_Δ), are different from zero in (a, ∞).

THEOREM 4.7 (Three-point boundary-value problem). Let $b(x, \lambda)$ have the property (v) for $x \in \langle a, \infty)$ and $\lambda \in (\Lambda_1, \Lambda_2)$ and let the hypotheses of the Oscillation Theorem 4.5 or those of Theorem 4.6 be met. Let $a \leqq b < c \in \langle a, \infty)$ be fixed numbers. Also, let $\alpha(\lambda)$, $\alpha_1(\lambda)$, $\beta(\lambda)$, $\beta_1(\lambda)$ be continuous functions of parameter $\lambda \in (\Lambda_1, \Lambda_2)$ with $|\alpha|+|\alpha_1| \neq 0$, $|\beta|+|\beta_1| \neq 0$, while either $\beta(\lambda) \equiv 0$ or $\beta(\lambda) \neq 0$ for all $\lambda \in (\Lambda_1, \Lambda_2)$.

Then there exist a positive integer v and a sequence of values of the parameter $\lambda \in (\Lambda_1, \Lambda_2)$ (eigenvalues)

$$\lambda_v, \lambda_{v+1}, \ldots, \lambda_{v+p}, \ldots$$

with a corresponding sequence of functions (eigenfunctions)

$$y_v, y_{v+1}, \ldots, y_{v+p}, \ldots$$

such that $y_{v+p} = y(x, \lambda_{v+p})$ is a solution of the differential equation (a_Δ) satisfying the boundary conditions

$$\begin{gathered} y(a, \lambda_{v+p}) = 0 \\ \alpha_1(\lambda_{v+p}) y(b, \lambda_{v+p}) - \alpha(\lambda_{v+p}) y'(b, \lambda_{v+p}) = 0\,, \\ \beta_1(\lambda_{v+p}) y(c, \lambda_{v+p}) - \beta(\lambda_{v+p}) y'(c, \lambda_{v+p}) = 0 \end{gathered} \tag{4.25}$$

and $y(x, \lambda_{v+p})$ has exactly $v+p$ zeros in (a, c).

PROOF. Let $y_1(x, \lambda)$, $y_2(x, \lambda)$ be solutions of (a_Δ) with $y_1(a, \lambda) = y_1'(a, \lambda) = 0$, $y_1''(a, \lambda) = 1$, $y_2(a, \lambda) = y_2''(a, \lambda) = 0$, $y_2'(a, \lambda) = 1$ for $\lambda \in (\Lambda_1, \Lambda_2)$. The set of solutions $y = c_1 y_1 + c_2 y_2$ forms a band of the first kind at a satisfying the second order equation (c) in which $w = w(x, \lambda) = y_1 y_2' - y_1' y_2$, $w(a, \lambda) = w'(a, \lambda) = 0$, $w''(a, \lambda) = -1$. In view of the assumption that $b(x, \lambda)$ has the property (v) in $\langle a, \infty)$ for $\lambda \in (\Lambda_1, \Lambda_2)$, we have $w(x, \lambda) < 0$ for $x > a$ and $\lambda \in (\Lambda_1, \Lambda_2)$, taking account of the integral identity (1.8) for the solution w of the differential equation

$$z''' + 2A(x, \lambda) z' + [A'(x, \lambda) - b(x, \lambda)] z = 0\,. \tag{b_Δ}$$

Thus the band $y = c_1 y_1 + c_2 y_2$ is regular in (a, ∞). Choose the constants c_1, c_2 to satisfy the second condition of (4.25). (The first one is obviously satisfied since $y(a, \lambda) = 0$ for $\lambda \in (\Lambda_1, \Lambda_2)$.) This is evidently possible, because if $a < b$ then it suffices to choose c_1, c_2 so that

$$\begin{aligned} c_1 y_1(b, \lambda) + c_2 y_2(b, \lambda) &= \alpha(\lambda)\,, \\ c_1 y_1'(b, \lambda) + c_2 y_2'(b, \lambda) &= \alpha_1(\lambda)\,. \end{aligned}$$

The determinant of this system is $w(x, \lambda) \neq 0$ for $x > a$ and $\lambda \in (\Lambda_1, \Lambda_2)$. If $a = b$, it is sufficient to put $c_1 = 1$, $c_2 = 0$, and $y(x, \lambda) = y_1(x, \lambda)$.

Denote by $y(x, \lambda)$ a solution of (a_Δ) satisfying the first two of the boundary conditions (4.25) for all $\lambda \in (\Lambda_1, \Lambda_2)$. Our aim now is to find $\lambda \in (\Lambda_1, \Lambda_2)$ for which $y(x, \lambda)$ satisfies also the third condition (4.25).

Let, for $\lambda = \bar{\lambda} \in (\Lambda_1, \Lambda_2)$, the solution $y(x, \lambda)$ have exactly ν zeros in (a, c). For the ν-th zero $x_\nu(\bar{\lambda})$ and the $(\nu+1)$-st zero $x_{\nu+1}(\bar{\lambda})$ of $y(x, \bar{\lambda})$ we have $x_\nu(\bar{\lambda}) < c \leqq x_{\nu+1}(\bar{\lambda})$.

It follows from the oscillation theorem that, for some $\bar{\bar{\lambda}} > \bar{\lambda} \in (\Lambda_1, \Lambda_2)$, we have $x_{\nu+1}(\bar{\bar{\lambda}}) < c$, while $x_{\nu+1}(\lambda) = c$ holds for no $\lambda \geqq \bar{\bar{\lambda}}$. Since $x_{\nu+1}(\lambda)$ is a continuous function of the parameter $\lambda \in (\Lambda_1, \Lambda_2)$ (Lemma 4.2), there exists a greatest $\bar{\lambda}^*_\nu \in (\bar{\lambda}, \bar{\bar{\lambda}})$ for which $y(c, \bar{\lambda}^*_\nu) = 0$ and $y(x, \bar{\lambda}^*_\nu)$ has exactly ν zeros in (a, c).

Thus $x_{\nu+1}(\bar{\lambda}^*_\nu) = c < x_{\nu+2}(\bar{\lambda}^*_\nu)$ for $\lambda = \bar{\lambda}^*_\nu$.

Also, the oscillation theorem implies that there exists $\lambda^* > \bar{\lambda}^*_\nu \in (\Lambda_1, \Lambda_2)$ such that for $\lambda = \lambda^*$ we have $x_{\nu+2}(\lambda^*) < c$, while $x_{\nu+2}(\lambda) = c$ holds for no $\lambda > \lambda^*$. From the continuity of $x_{\nu+2}(\lambda)$ with respect to the parameter $\lambda \in (\Lambda_1, \Lambda_2)$ (Lemma 4.2) it follows that in $(\bar{\lambda}^*_\nu, \lambda^*)$ there is a least $\bar{\lambda}_{\nu+1}$ and a greatest $\bar{\lambda}^*_{\nu+1}$ such that $y(c, \bar{\lambda}_{\nu+1}) = y(c, \bar{\lambda}^*_{\nu+1}) = 0$, while $y(x, \bar{\lambda}_{\nu+1})$ and $y(x, \bar{\lambda}^*_{\nu+1})$ have exactly $\nu+1$ zeros in (a, c).

Continuing in this manner, we can find a sequence of values of the parameter $\lambda \in (\Lambda_1, \Lambda_2)$,

$$\bar{\lambda}^*_\nu, \bar{\lambda}_{\nu+1}, \bar{\lambda}^*_{\nu+1}, \bar{\lambda}_{\nu+2}, \bar{\lambda}^*_{\nu+2}, \ldots, \bar{\lambda}_{\nu+p}, \bar{\lambda}^*_{\nu+p}, \ldots$$

to which there corresponds a sequence of functions

$$\bar{y}^*_\nu, \bar{y}_{\nu+1}, \bar{y}^*_{\nu+1}, \bar{y}_{\nu+2}, \bar{y}^*_{\nu+2}, \ldots, \bar{y}_{\nu+p}, \bar{y}^*_{\nu+p}, \ldots, p = 0, 1, \ldots$$

such that $\bar{y}_{\nu+p} = y(x, \bar{\lambda}_{\nu+p})$, $\bar{y}^*_{\nu+p} = y(x, \bar{\lambda}^*_{\nu+p})$ are solutions of (a_Δ) satisfying the first two of the conditions (4.25) and $y(c, \bar{\lambda}_{\nu+p}) = 0$, $y(c, \bar{\lambda}^*_{\nu+p}) = 0$, while $\bar{y}_{\nu+p}$ and $\bar{y}^*_{\nu+p}$ have exactly $\nu+p$ zeros in (a, c).

The theorem is immediately proved if we have $\beta(\lambda) \equiv 0$ for $\lambda \in (\Lambda_1, \Lambda_2)$. Suppose therefore that $\beta(\lambda) \neq 0$ for $\lambda \in (\Lambda_1, \Lambda_2)$. The ratio $\dfrac{y'(c, \lambda)}{y(c, \lambda)}$ for $\lambda \in (\bar{\lambda}^*_{\nu+p}, \bar{\lambda}_{\nu+p+1})$ ranges from $-\infty$ to ∞, because

$$\lim_{\lambda \to \bar{\lambda}^*_{\nu+p}} \frac{y'(c, \lambda)}{y(c, \lambda)} = +\infty, \quad \lim_{\lambda \to \bar{\lambda}_{\nu+p+1}} \frac{y'(c, \lambda)}{y(c, \lambda)} = -\infty .$$

Therefore, there is $\lambda_{\nu+p} \in (\bar{\lambda}^*_{\nu+p}, \bar{\lambda}_{\nu+p+1})$ such that

$$\frac{y'(c, \lambda_{\nu+p})}{y(c, \lambda_{\nu+p})} = \frac{\beta_1(\lambda_{\nu+p})}{\beta(\lambda_{\nu+p})}$$

and $y(x, \lambda_{\nu+p})$ has exactly $\nu+p$ zeros in (a, c). Thus the theorem is proved. □

Let y_1, y_2, y_3 be a fundamental system of solutions of the differential equation (a) with

$$y_1(a, \lambda) = y_1'(a, \lambda) = 0, \quad y_1''(a, \lambda) = 1,$$
$$y_2(a, \lambda) = y_2''(a, \lambda) = 0, \quad y_2'(a, \lambda) = 1,$$
$$y_3'(a, \lambda) = y_3''(a, \lambda) = 0, \quad y_3(a, \lambda) = 1.$$

The sets of solutions $y = c_1y_1 + c_2y_2$, $y = c_1y_1 + c_2y_3$, $y = c_1y_2 + c_2y_3$ form bands of the first, second, and third kind, respectively.

Let $w_1 = y_1y_2' - y_1'y_2$, $w_2 = y_1y_3' - y_1'y_3$, $w_3 = y_2y_3' - y_2'y_3$. The functions w_1, w_2, w_3 satisfy the differential equation (b_Δ). The function $w_1(x, \lambda) = w(x, \lambda)$, from the proof of Theorem 4.7. It is easy to check that $w_2(a, \lambda) = w_2''(a, \lambda) = 0, w_2'(a, \lambda) = -1, w_3(a, \lambda) = -1$, $w_3'(a, \lambda) = 0$, $w_3''(a, \lambda) = 2A(a, \lambda)$.

LEMMA 4.3. Let $w_1(x, \lambda) \neq 0$, $w_2(x, \lambda) \neq 0$, $w_3(x, \lambda) \neq 0$ for $x > a$ and $\lambda \in (\Lambda_1, \Lambda_2)$. If the solutions of the band of the first kind are oscillatory in (a, ∞), then so are also the solutions of the bands of the second and the third kinds of the differential equation (a_Δ).

The proof follows from the construction of bands and from the fact that whenever the band of the first kind oscillates in (a, ∞), then so do also the solutions y_1, y_2 while y_1 belongs, at the same time, to the band of the second kind and y_2 is in the band of the third kind. Moreover, the solutions in the bands of the first, second and third kind satisfy the differential equation (c), where $w = w_1$, $w = w_2$, $w = w_3$ at the respective bands. □

COROLLARY 4.4. If we assume, besides the hypotheses of Theorem 4.5 or 4.6, that $b(x, \lambda)$ has the property (v) in $\langle a, \infty)$ for $\lambda \in (\Lambda_1, \Lambda_2)$ and add the assumption $w_2(x, \lambda) \neq 0$ or $w_3(x, \lambda) \neq 0$ for $x > a$, then the assertion of the oscillation theorem remains valid also for bands of the second or of the third kind, respectively.

The assertion follows from Lemma 4.3, the rest of the proof being similar to that of Theorem 4.5 or 4.6, respectively.

REMARK 4.5. In order that $w_3(x, \lambda) \neq 0$ for $x > a$ and $\lambda \in (\Lambda_1, \Lambda_2)$, it is sufficient to assume, e.g. that $A(x, \lambda) \leqq 0$ for $x \in (a, \infty)$ and $\lambda \in (\Lambda_1, \Lambda_2)$, because then the integral identity (1.8) for $w_3(x, \lambda)$ reduces to

$$w_3(x, \lambda) w_3''(x, \lambda) - \frac{1}{2} w_3'^2(x, \lambda) + A(x, \lambda) w_3^2(x, \lambda) -$$

$$- \int_a^x b(t, \lambda) w_3^2(t, \lambda)\, dt = -A(a, \lambda)\,.$$

Consequently, $w_3(x, \lambda)$ cannot have a zero at $x > a$. As in Theorem 1.7, a sufficient condition in order that $w_2(x, \lambda) \neq 0$ for $x \in (a, \infty)$ and $\lambda \in (\Lambda_1, \Lambda_2)$ is $A(x, \lambda) \leqq 0$ and $A'(x, \lambda) + b(x, \lambda) \geqq 0$ for $x \in \langle a, \infty)$ and $\lambda \in (\Lambda_1, \Lambda_2)$.

COROLLARY 4.5. Corollary 4.4 implies the existence of eigenvalues and eigenfunctions for the boundary-value problem (a_Δ), (4.18), $i = 1, 2$.

A question arises, as to what conditions must be placed on the coefficients of the differential equation (a_Δ) so that the parameter v (in Theorem 4.7) should take the value 1. We shall try to answer this question.

The following lemma is easy to prove.

LEMMA 4.4. A third order linear homogeneous differential equation with constant coefficients is non-oscillatory in $\langle a, \infty)$ if and only if it is disconjugate in $\langle a, \infty)$.

Similarly, the Euler differential equation

$$z''' + \frac{p}{x^2} z' + \frac{\frac{\varepsilon}{2} - p}{x^3} z = 0$$

is non-oscillatory in $(0, \infty)$ if and only if it is disconjugate in $(0, \infty)$.

To simplify our further work, let $a > 0$.

THEOREM 4.8. Let the coefficients of the differential equation (a_Δ) satisfy either

$$A(x, \lambda) \leqq \frac{p}{2}, \quad 0 \leqq b(x, \lambda) \leqq \frac{q}{2} \tag{4.26}$$

for $x \in \langle a, \infty)$ and $\lambda \in (\Lambda_1, \Lambda_2)$, where

$$p \leqq 0, \quad q \leqq \frac{4}{3\sqrt{3}}(-p)^{3/2}$$

are constants, or

$$A(x, \lambda) \leqq \frac{p}{2x^2}, \quad 0 \leqq b(x, \lambda) \leqq \frac{\varepsilon}{2x^3} \tag{4.27}$$

for $x \in \langle a, \infty)$ and $\lambda \in (\Lambda_1, \Lambda_2)$, where

$$p \leqq 1, \quad \varepsilon \leqq \frac{4}{3\sqrt{3}}(1-p)^{3/2},$$

p, ε being constants. Then the differential equation (a_Δ) is disconjugate in $\langle a, \infty)$.

PROOF. It follows from Theorem 2.63 that under the assumptions (4.26) or (4.27) the differential equation (a_Δ) is non-oscillatory in $\langle a, \infty)$ for $\lambda \in (\Lambda_1, \Lambda_2)$. The relations between p and q or between p and ε imply respectively, that (4.19) or the given Euler differential equation, is non-oscillatory and disconjugate by Lemma 4.4.

Let $\bar{a} \geqq a$. Let $y(x, \lambda)$ be a solution of (a_Δ) with $y(\bar{a}, \lambda) = y'(\bar{a}, \lambda) = 0$, $y''(\bar{a}, \lambda) \neq 0$. We show that this solution has no other zeros in $\langle a, \infty)$. In fact, if we assume that $y(\bar{x}, \lambda) = 0$, $\bar{x} > \bar{a}$, then by Theorem 2.57 every solution of (4.19) or of the Euler equation, respectively, with a simple zero at $\bar{a}$ has another zero in $(\bar{a}, \bar{x}\rangle$, which is a contradiction. For $\bar{x} < \bar{a}$, the proof is obvious.

Identity (4.18) implies that every solution of (a_Δ) having a simple zero at $\bar{a}$ has at most one other zero in $\langle a, \infty)$. Thus the theorem is proved. □

THEOREM 4.9. Let the function $b(x, \lambda)$ have the property (v) in $\langle a, \infty)$ for $\lambda \in (\Lambda_1, \Lambda_2)$. Also, assume

a) there exist numbers K_1, K_2 such that

$$K_1 \leqq 2A(x, \lambda) \leqq K_2 < 0 \text{ for } x \in \langle a, \infty) \text{ and } \lambda \in (\Lambda_1, \Lambda_2)$$

and

$$\lim_{\lambda \to \Lambda_2} b(x, \lambda) = \infty \text{ uniformly for } x \in \langle a, \infty)$$

or

b) there is a number $p < 0$ with

$$A(x, \lambda) \leqq \frac{p}{2x^2} \text{ for } x \in \langle a, \infty) \text{ and } \lambda \in (\Lambda_1, \Lambda_2)$$

and

$$\lim_{\lambda \to \Lambda_2} b(x, \lambda) = \infty \text{ uniformly for } x \in \langle a, \infty)$$

and

$$\lim_{\lambda \to \Lambda_1} b(x, \lambda) x^3 = 0 \text{ uniformly in } \langle a, \infty)$$

or

c) $\lim_{\lambda \to \Lambda_2} r(\lambda) = \infty$, where $r(\lambda)$ is given by (4.23)

and

$$2A(x, \lambda) \leqq K_2 < 0 \text{ for } x \in \langle a, \infty) \text{ and } \lambda \in (\Lambda_1, \Lambda_2)$$

and

$$\lim_{\lambda \to \Lambda_1} b(x, \lambda) = 0 \text{ uniformly for } x \in \langle a, \infty).$$

Then the assertion of Theorem 4.7 for $\nu = 1$ holds true.

PROOF. The hypotheses of Theorem 4.7 being met, the assertion of Theorem 4.7 holds true. We have to show that $\nu = 0$ if $\beta(\lambda) \equiv 0$, and $\nu = 1$ if $\beta(\lambda) \neq 0$ for $\lambda \in (\Lambda_1, \Lambda_2)$.

Assume first that a) or c) are satisfied. Since $\lim_{\lambda \to \Lambda_1} b(x, \lambda) = 0$ uniformly in $\langle a, \infty)$, for $\varepsilon = (4/3)^{3/2}(-K_2)^{3/2}$ there exists a λ_0 such that

$$0 \leqq b(x, \lambda_0) < \frac{2}{3\sqrt{3}}(-K_2)^{3/2},$$

where K_2 is a number with $2A(x, \lambda) \leqq K_2$. It follows from Theorem 4.8 that the differential equation (a_Δ) is disconjugate in $\langle a, \infty)$ for $\lambda = \lambda_0$. Since $y(x, \lambda_0)$ is a solution of (a_Δ) vanishing at a, this solution has at most one other zero in (a, c). The zeros of $y(x, \lambda)$ being continuous functions of the parameter $\lambda \in (\Lambda_1, \Lambda_2)$, from the assumption that $\lim_{\lambda \to \Lambda_2} b(x, \lambda) = \infty$ uniformly in $\langle a, \infty)$ there follows the existence of $\bar{\lambda}$ such that the solution $y(x, \bar{\lambda})$ has exactly one zero in (a, c).

Let the assumptions c) be satisfied. As $\lim_{\lambda \to \Lambda_1} b(x, \lambda)x^3 = 0$ uniformly in $\langle a, \infty)$, we deduce that for $\varepsilon > 0$ with $\varepsilon < (4/3)^{3/2}(1-p)^{3/2}$, there exists λ_0 such that $0 \leqq b(x, \lambda_0)x^3 \leqq \varepsilon/2$ for all $x \in \langle a, \infty)$, that is, $0 \leqq b(x, \lambda_0) \leqq \varepsilon/2x^3$, where $\varepsilon \leqq (4/3)^{3/2}(1-p)^{3/2}$. Theorem 4.8 implies that the differential equation (a_Δ) is disconjugate in $\langle a, \infty)$ for $\lambda = \lambda_0$. Due to the condition $\lim_{\lambda \to \Lambda_2} b(x, \lambda) = \infty$ uniformly for $x \in \langle a, \infty)$, there exists $\bar{\lambda}$ such that the solution $y(x, \bar{\lambda})$ of (a_Δ) vanishing at a has exactly one other zero in (a, c).

Thus in each of the cases a), b), c) there is a $\bar{\lambda}$ such that the solution $y(x, \bar{\lambda})$ having a zero at a has exactly one zero in (a, c). Continuing as in the proof of Theorem 4.7, we obtain the assertion of the theorem. □

4. *Special Boundary-value Problems*

We begin by presenting two specially formulated boundary-value problems for a third order differential equation with constant coefficients, solved by Sansone [125], and continue with further results concerning the equation with constant coefficients (Greguš [42]). Then we extend the results to a differential equation with variable coefficients (Greguš [47, 55]) and finally solve a three-point boundary-value problem for an equation of the form (a) whose coefficients depend on two parameters (Greguš [53, 54]).

Consider the following third order differential equation with constant coefficients

$$y''' + 2A(\lambda)y' + \Omega(\lambda)y = 0\,, \tag{4.28}$$

where $A(\lambda)$ and $\Omega(\lambda)$ are constants depending continuously on the parameter $\lambda \in (\Lambda_0, \infty)$.

The differential equation adjoint to (4.28) is

$$z''' + 2A(\lambda)z' - \Omega(\lambda)z = 0\,. \tag{4.29}$$

Sansone [125] has proved the following result.

Let $A(\lambda) > 0$, $\Omega(\lambda) > 0$ for $\lambda \in (\Lambda_0, \infty)$ and let $\lim_{\lambda\to\infty} A(\lambda) = \infty$ and, moreover, let

$$\frac{\Omega(\lambda)}{[A(\lambda)]^{3/2}} \leqq \frac{2 \ln 3}{9\pi}\,.$$

Then for any $\delta > 0$ there exists $\lambda_0 \in (\Lambda_0, \infty)$ such that for every $\lambda > \lambda_0$ the set of solutions $y(x, \lambda)$ of (4.28) with $y'(x_0, \lambda) = 0$, $-\infty < x_0 < \infty$, contains one non-trivial solution (up to linear dependence) which vanishes together with its first derivative at some point of $(x_0, x_0 + \delta)$.

The proof is effected by examining the general solution of the differential equation (4.28) satisfying the required boundary conditions. The calculations are tedious, so we do not give the proof here, however, we shall give the proof of the following theorem.

THEOREM 4.10. Let $A(\lambda) > 0$, $\Omega(\lambda) > 0$ and let $\Omega(\lambda)/A(\lambda) \leqq L$ for $\lambda > \Lambda_0$, where $L > 0$ is a constant, independent of λ, and moreover let $\lim_{\lambda\to\infty} A(\lambda) = \infty$. Let $-\infty < x_0 < \infty$ and let $d > 0$ be a real number. Then there exist infinitely many values of the parameter $\lambda \in (\Lambda_0, \infty)$ (eigenvalues) with the corresponding set of non-trivial solutions $y(x, \lambda)$ (eigenfunctions) of the differential equation (4.28) which satisfy the boundary conditions

$$y'(x_0, \lambda) = y(x_0 + d, \lambda) = y'(x_0 + d, \lambda) = 0\,. \tag{4.30}$$

PROOF. The characteristic equation

$$\varrho^3 + 2A(\lambda)\varrho + \Omega(\lambda) = 0 \tag{4.31}$$

has no positive real root. However, the characteristic equation of the adjoint equation,

$$\varrho^3+2A(\lambda)\varrho-\Omega(\lambda)=0\,, \tag{4.32}$$

does have a positive real root, denote it by $\alpha(\lambda)$. The roots of (4.31) have the form

$$-\alpha(\lambda),\ \frac{\alpha(\lambda)}{2}+i\beta(\lambda),\ \frac{\alpha(\lambda)}{2}-i\beta(\lambda),\ i=\sqrt{-1}\,,$$

where $\alpha(\lambda)>0$, $\beta(\lambda)>0$ for $\lambda>\Lambda_0$.

Solving (4.32) gives $\alpha(\lambda)$ in the form

$$\alpha(\lambda)=\sqrt[3]{\left\{\frac{\Omega(\lambda)}{2}+\sqrt{\frac{\Omega^2(\lambda)}{4}-\frac{8A^3(\lambda)}{27}}\right\}}+$$

$$+\sqrt[3]{\left\{\frac{\Omega(\lambda)}{2}-\sqrt{\frac{\Omega^2(\lambda)}{4}+\frac{8A^3(\lambda)}{27}}\right\}}=$$

$$=\frac{\Omega(\lambda)}{A(\lambda)}\left[\sqrt[3]{\left\{\frac{\Omega(\lambda)}{2A^{3/2}(\lambda)}+\sqrt{\frac{\Omega^2(\lambda)}{4A^3(\lambda)}+\frac{8}{27}}\right\}}+\frac{2}{3}+\right.$$

$$\left.+\sqrt[3]{\left\{\frac{\Omega(\lambda)}{2A(\lambda)^{3/2}}-\sqrt{\frac{\Omega^2(\lambda)}{4A^3(\lambda)}+\frac{8}{27}}\right\}^2}\right]^{-1}.$$

The expression in brackets [ ] has limit 3/2 as $\lambda\to\infty$, and from the assumption $\Omega(\lambda)/A(\lambda)\leqq L$, it follows, that $\alpha(\lambda)$ is bounded.

From the requirement that for some $\lambda\in(\Lambda_0,\infty)$ the solution $y(x,\lambda)$ should satisfy the boundary conditions (4.30), we obtain the following equations for the constants c_1, c_2, c_3 in the general solution:

$$-c_1\alpha+c_2\frac{\alpha}{2}+c_3\beta=0\,,$$

$$c_1\,e^{-d\alpha}+c_2\,e^{d\alpha/2}\cos d\beta+c_3\,e^{d\alpha/2}\sin d\beta=0\,,$$

$$-c_1\alpha\,e^{-d\alpha}+c_2\frac{\alpha}{2}\,e^{d\alpha/2}\cos d\beta+c_3\frac{\alpha}{2}\,e^{d\alpha/2}\sin d\beta-$$

$$-c_2\beta\, e^{d\alpha/2}\sin d\beta + c_3\beta\, e^{d\alpha/2}\cos d\beta = 0\ ,$$

whence it follows that λ must be such that the equation

$$\begin{vmatrix} -\alpha, & \dfrac{\alpha}{2}, & \beta \\ e^{-d\alpha}, & e^{d\alpha/2}, & e^{d\alpha/2}\sin d\beta \\ -\alpha\, e^{-d\alpha}, & \dfrac{\alpha}{2}e^{d\alpha/2}\cos d\beta - \beta\, e^{d\alpha/2}\sin d\beta, & \dfrac{\alpha}{2}e^{d\alpha/2}\sin d\beta + \\ & & +\beta\, e^{d\alpha/2}\cos d\beta \end{vmatrix} = 0$$

holds. Multiplying this last equation by $-(\alpha\beta)^{-1}\exp(2^{-1}\,d\alpha)$ we get

$$e^{(3/2)d} + \frac{\frac{3}{4}\alpha^2+\beta^2}{\alpha\beta}\sin d\beta - \cos d\beta = 0\ . \tag{4.33}$$

The expression $\exp((3/2)d\alpha)$ is bounded and

$$\lim_{\lambda\to\infty}\left(\frac{3}{4}\alpha^2+\beta^2\right)\frac{1}{\alpha\beta} = \infty, \quad \lim_{\lambda\to\infty} d\beta = \infty\ .$$

We conclude that (4.33) has roots for infinitely many values of λ, so the theorem is proved.□

LEMMA 4.5. A necessary and sufficient condition for a non-trivial solution y of the differential equation (4.28) to satisfy $y'(x_0,\lambda) = y(x_1,\lambda) = y'(x_1,\lambda) = 0$ is that any solution z of (4.29) with $z(x_0,\lambda) = z''(x_0,\lambda) = 0$ should vanish at x_1, where $x_0 \neq x_1 \in (-\infty,\infty)$.

The proof follows from Theorem 1.4.

LEMMA 4.6. Let $y(x,\lambda)$ be a solution of the differential equation (4.28). Then the function $y(x_0-(x-x_0),\lambda) = z(x,\lambda)$ satisfies the differential equation (4.29), for $-\infty < x_0 < \infty$.

PROOF. Let $y(x,\lambda)$ be a solution of (4.28). Construct $z(x,\lambda) \stackrel{d}{=} y(x_0-(x-x_0),\lambda)$ and substitute in (4.29). We get

$$\begin{aligned} &-y'''(x_0-(x-x_0),\lambda) - 2A(\lambda)y'(x_0-(x-x_0),\lambda) - \\ &-\Omega(\lambda)y(x_0-(x-x_0),\lambda) = -\{y'''(x_0-(x-x_0),\lambda) + \\ &+2A(\lambda)y'(x_0-(x-x_0),\lambda) + \Omega(\lambda)y(x_0-(x-x_0),\lambda)\} = 0\ , \end{aligned}$$

since the expression in braces equals zero. Thus the lemma is proved. □

REMARK 4.6. Let $y(x, \lambda)$ be a solution of the differential equation (4.28) with $y(x_0, \lambda) = y''(x_0, \lambda) = 0$. Then $\int_{x_0}^{x} y(t, \lambda)\, dt$ is also a solution of (4.28) with a double zero at x_0.

The proof is carried out by substitution in (4.28) and termwise differentiation with respect to x.

REMARK 4.7. Theorem 4.10 may be restated for a solution of the differential equation (4.29) as follows.

Let $A(\lambda)$, $\Omega(\lambda)$ satisfy the hypotheses of Theorem 4.10. Let $d > 0$ and let $x_0 - d$ and x_0 be two fixed points, $-\infty < x_0 < \infty$. Then there exist infinitely many values of the parameter $\lambda \in (\Lambda_0, \infty)$ (eigenvalues) to which there correspond eigenfunctions $z(x, \lambda)$ solving (4.29) and satisfying the boundary conditions

$$z(x_0 - d, \lambda) = z'(x_0 - d, \lambda) = z'(x_0, \lambda) = 0 .$$

The proof follows directly from Lemma 4.6.

THEOREM 4.11. Let the hypotheses of Theorem 4.10 be satisfied. Then there exist infinitely many values of the parameter $\lambda \in (\Lambda_0, \infty)$ (eigenvalues) to which there correspond eigenfunctions $y(x, \lambda)$ which are solutions of the differential equation (4.28) and satisfy the boundary conditions

$$y(x_0 - d, \lambda) = y(x_0, \lambda) = y''(x_0, \lambda) = 0,$$
$$-\infty < x_0 < \infty, \ d > 0 .$$

The proof follows from Theorem 4.10, Lemma 4.5 and Lemma 4.6. □

LEMMA 4.7. Let $A(x, \lambda) > 0$, $A'(x, \lambda) \leq 0$ be continuous functions of $x \in \langle a, \infty)$ and $\lambda \in (\Lambda_0, \infty)$ and assume that $A'(x, \lambda) \not\equiv 0$ in any subinterval of $\langle a, \infty)$ for $\lambda > \Lambda_0$. Also, let $\lim_{\lambda \to \infty} A(x, \lambda) = \infty$ uniformly for all $x \in \langle a, \infty)$. Let $\delta > 0$ be a real number. Then there exist infinitely many values of the parameter $\lambda \in (\Lambda_0, \infty)$ (eigenvalues) with the corresponding solutions $y(x, \lambda)$ (eigenfunctions) of the differential equation

$$y''' + 2A(x, \lambda)y' = 0 \tag{4.34}$$

satisfying the boundary conditions

$$y'(a, \lambda) = 0,\ y(a+\delta, \lambda) = y'(a+\delta, \lambda) = 0\ .$$

PROOF. The adjoint differential equation to (4.34) is

$$z''' + 2A(x, \lambda)z' + 2A'(x, \lambda)z = 0\ , \tag{4.35}$$

which is obtained by differentiating the second order differential equation

$$z'' + 2A(x, \lambda)z = 0\ . \tag{4.36}$$

Let $z(x, \lambda)$ be a non-trivial solution of the differential equation (4.36) with $z(a, \lambda) = 0$. Then $z''(a, \lambda) = 0$. At the same time, $z(x, \lambda)$ is a solution of (4.35). The Sturm oscillation theorem for second order differential equations (Sansone [124]) implies the existence of infinitely many values of the parameter $\lambda \in (\Lambda_0, \infty)$ (eigenvalues) and of the corresponding solutions $z(x, \lambda)$ of (4.36) which at the same time solve (4.35) and also satisfy $z(a, \lambda) = z''(a, \lambda) = z(a+\delta, \lambda) = 0$. By Theorem 1.4 there exists, for any such eigenvalue λ, a solution $y(x, \lambda)$ of (4.34) satisfying the boundary conditions

$$y'(a, \lambda) = y(a+\delta, \lambda) = y'(a+\delta, \lambda) = 0\ .$$

Thus the lemma is proved. □

COROLLARY 4.6. For every eigenvalue mentioned in Lemma 4.7 there is a solution $v(x, \lambda)$ of the differential equation (4.35) which satisfies the boundary conditions

$$v(a, \lambda) = v''(a, \lambda) = 0,\ \ v(a+\delta, \lambda) = 0\ .$$

The proof follows from Theorem 1.4.

LEMMA 4.8. Suppose that the hypotheses of Lemma 4.7 are fulfilled and let $b(x, \lambda)$ have the property (v) in $\langle a, \infty)$ for all $\lambda > \Lambda_0$. Also, let $A'(x, \lambda) + b(x, \lambda) \leqq 0$ for every $x \in \langle a, \infty)$ and $\lambda > \Lambda_0$. Then, given any eigenvalue from Lemma 4.7, there is a solution $z(x, \lambda)$ of the differential equation (b_Δ) satisfying $z(a, \lambda) = z''(a, \lambda) = 0$, $z'(a, \lambda) \neq 0$ and having another zero in $(a, a+\delta)$, $\delta > 0$ from Lemma 4.7.

PROOF. The differential equation (b_Δ) may be written in the following form:

$$z''' + 2A(x, \lambda)z' + 2A'(x, \lambda)z = [A'(x, \lambda) + b(x, \lambda)]z .$$

Let λ^* be one of the eigenvalues from Lemma 4.7 and let $v(x, \lambda^*)$ be the corresponding eigenfunction, i.e. the solution of (4.35) meeting the boundary conditions $v(a, \lambda^*) = v''(a, \lambda^*) = 0$, $v'(a, \lambda^*) = k$, $v(a+\delta, \lambda^*) = 0$, k being a positive constant. Let $z(x, \lambda^*)$ be a solution of (b_Δ) with $z(a, \lambda^*) = z''(a, \lambda^*) = 0$, $z'(a, \lambda^*) = k$. Let v_1, v_2, v_3 be a fundamental set of solutions of (4.35) whose wronskian equals one. By variation of constants we deduce that the solution $z(x, \lambda^*)$ can be written in the form

$$z(x, \lambda^*) = v(x, \lambda^*) + \\ + \int_a^x [A'(t, \lambda^*) + b(t, \lambda^*)] z(t, \lambda^*) W(x, t)\, dt, \qquad (4.37)$$

where

$$W(x, t) = \begin{vmatrix} v_1(x, \lambda^*), & v_2(x, \lambda^*), & v_3(x, \lambda^*) \\ v_1(t, \lambda^*), & v_2(t, \lambda^*), & v_3(t, \lambda^*) \\ v_1'(t, \lambda^*), & v_2'(t, \lambda^*), & v_3'(t, \lambda) \end{vmatrix} .$$

For a given t, $W(x, t) = \bar{v}(x, \lambda^*)$ is a solution of (4.35) with $\bar{v}(t, \lambda^*) = \bar{v}'(t, \lambda^*) = 0$, $\bar{v}''(t, \lambda^*) = 1$. It follows from (1.6) that $\bar{v}(x, \lambda^*) \geqq 0$ for $t \leqq x$, that is, $W(x, t) \geqq 0$ for $t \leqq x$.

The assumption that $z(x, \lambda^*)$ has no zero in $(a, a+\delta)$ contradicts (4.37). Thus the lemma is proved. □

THEOREM 4.12. Let the coefficients of the differential equation (a_Δ) satisfy the hypotheses of Lemma 4.8. Then there exist infinitely many values of the parameter $\lambda \in (\Lambda_0, \infty)$ with corresponding non-trivial solutions $y(x, \lambda)$ of the differential equation (a_Δ) satisfying $y'(a, \lambda)=0$ and having a double zero in $(a, a+\delta)$.

The proof follows from Lemma 4.8 and Theorem 1.4. □

REMARK 4.8. Some further boundary-value problems are formulated and solved in papers by Greguš [40, 42, 49].

Let us now consider the following third order differential equation

$$y'''+2A(x, \lambda, \mu)y'+[A'(x, \lambda, \mu)+b(x, \lambda, \mu)]y=0, \tag{a_{λ_μ}}$$

where $A(x, \lambda, \mu)$, $A'(x, \lambda, \mu)=(d/dx)A(x, \lambda, \mu)$, $b(x, \lambda, \mu)$ are continuous functions of $x\in\langle a, c\rangle$, $\lambda\in(\Lambda_1, \Lambda_2)$, $\mu\in(M_1, M_2)$.

Our aim is to establish sufficient conditions on the coefficients of the differential equation (a_{λ_μ}) in order that parameters λ, μ can be chosen so that there exists a non-trivial solution y of (a_{λ_μ}) satisfying the boundary conditions

$$y(a, \lambda, \mu)=y'(a, \lambda, \mu)=y(b, \lambda, \mu)=y(c, \lambda, \mu)=0\,, \tag{4.38}$$

where $a<b<c$. Originally, the problem was formulated and solved for a second order differential equation (Greguš, et al. [60]).

THEOREM 4.13. *Let $a<b<c$ be real numbers. Let $A(x, \lambda)\geqq k$, $A'(x, \lambda)=(d/dx)A(x, \lambda)$ be continuous functions of $x\in\langle a, c\rangle$ and $\lambda\in(\Lambda_1, \Lambda_2)$. Let $b(x, \lambda, \mu)$ be of the form*

$$b(x, \lambda, \mu)=b_\lambda(x, \lambda)+b_\mu(x, \mu)\,,$$

where $b_\lambda(x, \lambda)$ is a continuous function of $x\in\langle a, c\rangle$ and $\lambda\in(\Lambda_1, \Lambda_2)$ and

$$b_\mu(x, \mu)=\begin{cases} r(x) & \text{for } x\in\langle a, b),\\ s(x, \mu) & \text{for } x\in\langle b, c\rangle,\ \mu\in(M_1, M_2),\end{cases}$$

where r, s are assumed to be continuous functions of x and μ in the respective intervals, satisfying $s(b, \mu)=r(b)$ for all $\mu\in(M_1, M_2)$. Moreover, let $b(x, \lambda, \mu)$ be a non-negative function for $x\in\langle a, c\rangle$,

$\lambda \in (\Lambda_1, \Lambda_2)$ and $\mu \in (M_1, M_2)$, and let $\lim_{\lambda \to \Lambda_2} b_\lambda(x, \lambda) = \infty$ uniformly for $x \in \langle \beta, c \rangle$, where $b < \beta < c$.

Then there exist an integer N and sequences $\left\{\lambda_{N+p}\right\}_{p=0}^{\infty}$, $\left\{\mu_{N+p}\right\}_{p=0}^{\infty}$, $\left\{y_{N+p}\right\}_{p=0}^{\infty}$, $\lambda_{N+p} \in (\Lambda_1, \Lambda_2)$, $\mu_{N+p} \in (M_1, M_2)$, $p = 0, 1, \ldots$, such that $y_{N+p} = y(x, \lambda_{N+p}, \mu_{N+p})$, $p = 0, 1, \ldots$, is a solution of (a_{λ_μ}), satisfies (4.38) and has exactly $N+p$ zeros in $\langle a, b \rangle$.

PROOF. To begin with, extend the domains of the functions $A(x, \lambda)$, $A'(x, \lambda)$, $b(x, \lambda, \mu)$ to $\langle c, \infty) \times (\Lambda_1, \Lambda_2)$ or $\langle c, \infty) \times (\Lambda_1, \Lambda_2) \times (M_1, M_2)$, respectively, defining

$$A(x, \lambda) \stackrel{d}{=} A(c, \lambda), \quad A'(x, \lambda) \stackrel{d}{=} A'(c, \lambda), \quad b(x, \lambda, \mu) \stackrel{d}{=}$$
$$\stackrel{d}{=} b(c, \lambda, \mu) \quad \text{for } x \geqq c, \ \lambda \in (\Lambda_1, \Lambda_2), \ \mu \in (M_1, M_2) .$$

Let $y(x, \lambda, \mu)$ be a solution of (a_{λ_μ}) with $y(a, \lambda, \mu) = y'(a, \lambda, \mu) = 0$, $y''(a, \lambda, \mu) \neq 0$ for $\lambda \in (\Lambda_1, \Lambda_2)$, $\mu \in (M_1, M_2)$. Oscillation Theorem 4.5 implies that for some $\bar{\lambda}$ and $\bar{\mu}$ the solution $y(x, \bar{\lambda}, \bar{\mu})$ has exactly N zeros in (a, b), where N is a positive integer. Evidently,

$$x_N(\bar{\lambda}, \bar{\mu}) < b \leqq x_{N+1}(\bar{\lambda}, \bar{\mu}) ,$$

x_N being the N-th zero of $y(x, \bar{\lambda}, \bar{\mu})$ in (a, b).

Also, it follows from Theorem 4.5 that there exists $\lambda^* \in \langle \bar{\lambda}, \Lambda_2)$ with $x_{N+1}(\lambda^*, \bar{\mu}) < b$. By Lemma 4.2, there exists also $\lambda_N \in \langle \bar{\lambda}, \lambda^*)$ such that $y(b, \lambda_N, \bar{\mu}) = 0$ and $y(x, \lambda_N, \bar{\mu})$ has exactly N zeros in (a, b).

For $\lambda = \lambda_N$ and $\mu = \bar{\mu}$, let $y(x, \lambda_N, \bar{\mu})$ have exactly ν zeros in (b, c), where ν is a positive integer. Evidently,

$$\xi_\nu(\lambda_N, \bar{\mu}) < c < \xi_{\nu+1}(\lambda_N, \bar{\mu}) ,$$

ξ_ν being the ν-th zero of $y(x, \lambda_N, \bar{\mu})$ in (b, c). The oscillation theorem implies the existence of $\mu^* \in (\bar{\mu}, M_2)$ such that $\xi_{\nu+1}(\lambda_N, \mu^*) < c$.

Then from Lemma 4.2 there follows the existence of $\mu_N \in \langle \bar{\mu}, \mu^*)$ with $y(c, \lambda_N, \mu_N) = 0$. Putting $y_N = y(x, \lambda_N, \mu_N)$, we obtain the first terms of the sequences

$$\{\lambda_{N+p}\}_{p=0}^{\infty}, \quad \{\mu_{N+p}\}_{p=0}^{\infty}, \{y_{N+p}\}_{p=0}^{\infty} .$$

Proceeding in this way we establish the existence of further terms of these sequences. Thus the theorem is proved. □

THEOREM 4.14. Let $a<b<c$ be real numbers. Let $b(x, \lambda)$ be a continuous function of $y \in \langle a, c\rangle$ and $\lambda \in (\Lambda_1, \Lambda_2)$, and let it have the property (v) in $\langle a, c\rangle$ for $\lambda \in (\Lambda_1, \Lambda_2)$. Let $A(x, \lambda, \mu) \geqq 0$, $A'(x, \lambda, \mu) = (d/dx)A(x, \lambda, \mu)$ be continuous functions of $y \in \langle a, c\rangle$, $\lambda \in (\Lambda_1, \Lambda_2)$ and $\mu \in (M_1, M_2)$, and assume that $A(x, \lambda, \mu)$ has the form

$$A(x, \lambda, \mu) = A_\lambda(x, \lambda) + A_\mu(x, \mu) ,$$

where $A_\lambda(x, \lambda)$ and $A_\lambda'(x, \lambda) = (d/dx)A_\lambda(x, \lambda)$ are continuous functions of $x \in \langle a, c\rangle$ and $\lambda \in (\Lambda_1, \Lambda_2)$, and

$$A_\mu(x, \lambda) = \begin{cases} A_1(x) & \text{for } x \in \langle a, b\rangle, \\ A_2(x, \mu) & \text{for } x \in \langle b, c\rangle, \ \mu \in (M_1, M_2), \end{cases}$$

while $A_2(b, \mu) = A_1(b)$, $A_2'(b, \mu) = A_1'(b)$ for all $\mu \in (M_1, M_2)$. Let $A_\mu(x, \mu)$, $A_\mu'(x, \mu) = (d/dx)A_\mu(x, \mu)$ be continuous functions of $x \in \langle a, c\rangle$ and $\mu \in (M_1, M_2)$. Moreover, assume that $\lim_{\lambda \to \Lambda_2} A_\lambda(x, \lambda) = \infty$ uniformly for $x \in \langle a, c\rangle$ and that $\lim_{\mu \to M_2} A_\mu(x, \mu) = \infty$ uniformly for $x \in \langle \beta, c\rangle$, where $b<\beta<c$. Then there exist a positive integer N and sequences

$$\{\lambda_{N+p}\}_{p=0}^{\infty}, \ \{\mu_{N+p}\}_{p=0}^{\infty}, \ \{y_{N+p}\}_{p=0}^{\infty}$$

such that $y_{N+p} = y(x, \lambda_{N+p}, \mu_{N+p})$ is a solution of the differential equation (a_{λ_μ}) satisfying (4.38) and having exactly $N+p$ zeros in $\langle a, b\rangle$.

The proof of Theorem 4.14 is similar to that of Theorem 4.13.

Chapter II

Third Order Linear Homogeneous Differential Equations with Continuous Coefficients

§5. PRINCIPAL PROPERTIES OF SOLUTIONS OF LINEAR HOMOGENEOUS THIRD ORDER DIFFERENTIAL EQUATIONS WITH CONTINUOUS COEFFICIENTS

In this and the following sections of Chapter II we shall study the linear third order differential equation of the form

$$y''' + py'' + qy' + ry = 0 , \tag{A}$$

where $p = p(x)$, $q = q(x)$, $r = r(x)$ are continuous functions of $x \in (a, \infty)$, $a \geqq -\infty$. We shall deal especially with principles of the theory of bands and their applications and with results published in papers by Gera [26—35]. The extent of this monograph does not allow us to analyse in detail all the results; we shall therefore merely state some of them, referring to the relevant literature.

1. *Principal Properties of Solutions of the Differential Equation (A)*

By the differential equation adjoint to (A) we mean (Sansone [124])

$$[(z' - pz)' + qz]' - rz = 0 . \tag{B}$$

The differential equation (B) can be written as the following first order linear differential system:

$$\begin{aligned} z' &= -pz + u , \\ u' &= -qz + v , \\ v' &= rz , \end{aligned}$$

which implies (Sansone [124]) that for any $x_0 \in (a, \infty)$ and any triple of numbers z_0, z_0', z_0'' there exists exactly one solution of (B) defined on (a, ∞) and satisfying $z(x_0)=z_0$, $z'(x_0)=z_0'$, $(z'-pz)'(x_0)=z_0''$.

Let w be a solution of (B). Besides (A) and (B), consider also the differential equation

$$wy''+(pw-w')y'+(qw+(w'-pw)')y=0\,. \tag{C}$$

The operator on the left-hand side of (C) is a bilinear form resulting from the relationship between solutions of the mutually adjoint differential equations (A) and (B) (Sansone [124]).

Let us introduce the following notation for differential equations:

$$y''+py'+qy=0\,, \tag{A$_1$}$$

$$(z'-pz)'+qz=0\,, \tag{B$_1$}$$

$$y'''+py''+qy'=0\,, \tag{A$_2$}$$

$$[(z'-pz)'+qz]'=0\,. \tag{B$_2$}$$

LEMMA 5.1. Let the differential equation (B$_1$) be disconjugate in (a, ∞). If u is a solution of the differential equation (B$_1$) with $u(x_0)=0$, $u'(x_0)>0$, $a<x_0<\infty$, then $u(x)>0$ for $x>x_0$.

Proof is obvious. □

Let u_1, u_2 be a fundamental set of solutions of the differential equation (B$_1$). Then their wronskian is

$$W^0(x)=W^0(u_1,u_2)=\begin{vmatrix} u_1 & , \; u_2 \\ u_1'-pu_1, & u_2'-pu_2 \end{vmatrix}=\begin{vmatrix} u_1, & u_2 \\ u_1', & u_2' \end{vmatrix}\,.$$

The function

$$W_2^0(x,t)=\begin{vmatrix} u_1(x), & u_2(x) \\ u_1(t), & u_2(t) \end{vmatrix}$$

satisfies the differential equation (B$_1$) for any fixed $t\in(a,\infty)$. We have $W_2^0(t,t)=0$, $W_{2x}^0(t,t)\neq 0$. Consequently, $W_2^0(x,t)\neq 0$ for $x>t$ by Lemma 1.1.

LEMMA 5.2. Let the differential equation (B$_1$) be disconjugate in

(a, ∞). Let u_1, u_2 be a fundamental set of solutions of (B_1) and let $W_2^0(u_1, u_2) > 0$ for $x \in (a, \infty)$. The function

$$v(x) = -k \int_{x_0}^{x} \frac{W_2^0(x, t)}{W^0(t)} \, dt, \quad a < x_0 < \infty ,$$

where $k \neq 0$ is a constant, $x \in (a, \infty)$ is a solution of the differential equation (B_2) with $v(x_0) = 0$, $v'(x_0) = 0$, $(v' - pv)'(x_0) = k$, and $v(x) \neq 0$ for $x > x_0$.

PROOF. Using variation of constants, it is easy to verify that $v(x)$ satisfies the second order differential equation

$$(u' - pu)' + qu = k ,$$

and consequently also the equation (B_2). Evidently, $v(x_0) = 0$,

$$v'(x) = -k \int_{x_0}^{x} \frac{W_{2x}^0(x, t)}{W^0(t)} \, dt , \quad \text{hence} \quad v'(x_0) = 0 .$$

Also,

$$[v'(x) - p(x)v(x)]' =$$

$$= k - k \int_{x_0}^{x} \frac{1}{W^0(t)} \begin{vmatrix} [u_1'(x) - p(x)u_1(x)]', & [u_2'(x) - p(x)u_2(x)]' \\ u_1(t), & u_2(t) \end{vmatrix} dt ,$$

therefore $[v'(x) - p(x)v(x)]'(x_0) = k$. If $k \neq 0$, then $v(x) \neq 0$ for $x > x_0$. The lemma is proved. □

LEMMA 5.3. Let the differential equation (B_1) be disconjugate in (a, ∞) and let $r(x) \geqq 0$ for $x \in (a, \infty)$. If $w = w(x)$ is a solution of the differential equation (B) with $w(x_0) = w'(x_0) = 0$, $(w' - pw)'(x_0) = k > 0$, then $w(x) > 0$ for $x > x_0$, $a < x_0 < \infty$.

PROOF. The differential equation (B) may be written in the form

$$[(z' - pz)' + qz]' = rz .$$

The method of variation of constants yields that the solution w can be expressed as

$$w(x) = v(x) + \int_{x_0}^{x} \frac{W_3(x, t)}{W(t)} r(t)w(t)\,dt, \tag{5.1}$$

where v is a solution of (B_2) satisfying the same initial conditions at x_0 as the solution w, $W(t)$ is the wronskian of the fundamental system of solutions v_1, v_2, v_3 of (B_2), and

$$w_3(x, t) = \begin{vmatrix} v_1(x), & v_2(x), & v_3(x) \\ v_1(t), & v_2(t), & v_3(t) \\ v_1'(t), & v_2'(t), & v_3'(t) \end{vmatrix}.$$

If t is fixed, $W_3(x, t)$ is a solution of the differential equation (B_2) with a double zero at t. If $W(t) > 0$, it follows from Lemma 5.2 that $W_3(x, t) \geqq 0$ for $x \geqq t$. Then (5.1) implies the assertion of Lemma 5.3. □

REMARK 5.1. The relationship between the solutions of the differential equations (A) and (B) is analogous to that for mutually adjoint third order differential equations in the normal form (§1). We shall, therefore, not analyse it in more detail.

REMARK 5.2. The assumption that the differential equation (B_1) is disconjugate in (a, ∞) is equivalent to the assumption that the differential equation (A_1) is disconjugate in (a, ∞).

The assertion follows from the relationship between the solutions of adjoint second order differential equations (Sansone [124]). In fact, if y_1, y_2 form a fundamental set of solutions of the differential equation (A_1) with $y_1(x_0)y_2'(x_0) - y_1'(x_0)y_2(x_0) = 1$, then

$$z_1 = y_1(x) \exp\left(\int_{x_0}^{x} p(t)\,dt\right), \quad z_2 = y_2(x) \exp\left(\int_{x_0}^{x} p(t)\,dt\right)$$

is a fundamental set of solutions of the differential equation (B_1).

2. *Bands of Solutions of the Differential Equation (A)*

Let y_1, y_2, y_3 be a fundamental system of solutions of the differential equation (A) with

$$y_1(x_0)=y_1'(x_0)=0,\ \ y_1''(x_0)\neq 0,\ \ y_2(x_0)=y_2''(x_0)=0\,,$$
$$y_2'(x_0)\neq 0,\ \ y_3'(x_0)=y_3''(x_0)=0,\ \ y_3(x_0)\neq 0,\ \ x_0\in(a,\infty)\,. \tag{5.2}$$

The set of solutions $y=c_1y_1+c_2y_2$ of (A), where c_1, c_2 are arbitrary constants, is referred to as the band of solutions of the first kind at x_0.

Bands of the second and third kinds at x_0 are defined analogously, as in §1.

As we shall deal only with bands of the first kind at x_0, we shall say shortly "the band of solutions at x_0".

According to Remark 5.1, the function $w=y_1y_2'-y_1'y_2$ is a solution of the differential equation (B) with $w(x_0)=w'(x_0)=0$, $(w'-pw)'(x_0)\neq 0$.

It is easy to verify that the band of solutions at x_0 satisfies the differential equation (C).

COROLLARY 5.1. If the hypotheses of Lemma 5.3 are satisfied, we have $w(x)>0$ for $x>x_0$, and therefore the zeros of any two independent solutions from the band at x_0 separate each other in (x_0, ∞), i.e. the band of solutions of the differential equation (A) at x_0 is regular in (x_0, ∞).

REMARK 5.3. Let w be any solution of the differential equation (B) with $w(x)\neq 0$ for $x\in I\subset(a, \infty)$, where I is an interval. Then (by Remark 1.1 and Lemma 1.1) there exist two solutions $\bar{y}_1$, $\bar{y}_2$ of the differential equation (A) such that $w=\bar{y}_1\bar{y}_2'-\bar{y}_1'\bar{y}_2$ for $x\in I$. In the sequel, the set of solutions $y=c_1\bar{y}_1+c_2\bar{y}_2$ will be referred to as the band of solutions of the differential equation (A) in the interval I. The band in the interval I also satisfies the differential equation (C).

COROLLARY 5.2. Let the differential equation (B_1) be disconjugate in (a, ∞) and let $r(x)\geqq 0$ for $x\in(a, \infty)$. Let y_1 be a solution of the differential equation (A) such that $y_1(x_0)=y_1'(x_0)=0$, $y_1''(x_0)\neq 0$, $a<x_0<\infty$. Then $y_1(x)\neq 0$ for $x<x_0$.

PROOF. The assertion follows from the relationship between the solutions of adjoint differential equations (A) and (B) and from properties of bands. In fact, let y_1, y_2, y_3 be a fundamental set of solutions of (A) with condition (5.2) and let their wronskian be $W(y_1, y_2, y_3)(x_0)=1$. A fundamental system of solutions of the differential equation (B) has the form

$$z_1=(y_1y_2'-y_1'y_2)\,e^{\int_{x_0}^{x} p\,dt}, \quad z_2=(y_1y_3'-y_1'y_3)\,e^{\int_{x_0}^{x} p\,dt},$$

$$z_3=(y_2y_3'-y_2'y_3)\,e^{\int_{x_0}^{x} p\,dt}.$$

Clearly, $z_1(x_0)=z_1'(x_0)=0$, $(z_1'-pz_1)'(x_0) = -y_2'(x_0)y_1''(x_0)\neq 0$. Now suppose $y_1(x_1)=0$ for $a<x_1<x_0$. Then y_1 would be an element of the band at x_1, and could be written in the form $y_1(x)=c_1\bar{y}_1(x)+c_2\bar{y}_2(x)$, where $\bar{y}_1(x_1)=\bar{y}_1'(x_1)=0$, $\bar{y}_1''(x_1)\neq 0$, $\bar{y}_2(x_1)=\bar{y}_2''(x_1)=0$, $\bar{y}_2'(x_1)\neq 0$. Consequently, there would exist constants c_1, c_2 (at least one of them being non-zero) such that

$$c_1\bar{y}_1(x_0)+c_2\bar{y}_2(x_0)=0,$$

$$c_1\bar{y}_1'(x_0)+c_2\bar{y}_2'(x_0)=0.$$

This, however, contradicts the assertion of Lemma 5.3 and the regularity of the band at x_1 in the interval (x_1, ∞). Thus the proof is complete. □

LEMMA 5.4. Let the hypotheses of Corollary 5.2 be met and let $x_1>x_0$ be the first zero of y_1 (which has a double zero at x_0) to the right of x_0. Then every solution of the band at x_0 which is independent of y_1 has exactly one zero between x_0 and x_1.

The proof of Lemma 5.4 is similar to that of Theorem 1.8 and is therefore omitted.

LEMMA 5.5. Let the hypotheses of Lemma 5.4 be satisfied and let $x_1>x_0$ be the first zero of y_1 (having a double zero at x_0) to the right of x_0. Let $x_0<\bar{x}<x_1$ and let $\bar{y}_1$ be a solution of the differential equation (A) having a double zero at $\bar{x}$ and another zero at $\bar{x}_1>\bar{x}$. Then $\bar{x}_1>x_1$.

PROOF. Suppose the contrary, i.e. that $x_0<\bar{x}<\bar{x}_1\leqq x_1$. From the point x_0 there emerges a solution y_{x_0} belonging to the band at x_0 and passing through $\bar{x}$, thus belonging also to the band at $\bar{x}$. By Lemma 5.4, it necessarily has another zero in the interval $(\bar{x}, x_1\rangle$. Thus y_{x_0} would have two zeros in $(x_0, x_1\rangle$, which is impossible. The lemma is proved. □

3. Application of Bands to Solving a Three-point Boundary-value problem

Consider the differential equation

$$y''' + p(x)y'' + q(x)y' + \lambda r(x)y = 0\,, \tag{A_λ}$$

where p, q, r are continuous functions of $x \in \langle a, \infty)$, $r(x) > 0$ for $x \geqq a$, and $\lambda \geqq 0$ is a parameter. Moreover, let the differential equation (A_1) be disconjugate in $\langle a, \infty)$.

LEMMA 5.6. Let the above assumptions be met and let $a < b < \infty$. Assume that $y_1(x, \lambda)$ is a solution of the differential equation (A_λ) with $y_1(a, \lambda) = y_1'(a, \lambda) = 0$, $y_1''(a, \lambda) = k \neq 0$. Then there exists $\bar{\lambda} > 0$ such that, for every $\lambda > \bar{\lambda}$, the solution $y_1(x, \lambda)$ has at least one zero in (a, b).

PROOF. Compare the differential equation (A_λ) with

$$v''' + p(x)v'' + q(x)v' + \lambda kv = 0\,, \tag{$\bar{A}_\lambda$}$$

where $k = \min r(x)$ over $x \in \langle a, b\rangle$. The differential equation (A_λ) can be written in the form

$$y''' + p(x)y'' + q(x)y' + \lambda ky = [\lambda k - \lambda r(x)]y\,.$$

By the method of variation of constants for $y_1(x, \lambda)$ we get

$$y_1(x, \lambda) = \bar{v}_1(x, \lambda) - \lambda \int_a^x [r(x) - k]W(x, t)\, e^{\int_a^t p(s)\, ds}\, y_1(t, \lambda)\, dt\,, \tag{5.3}$$

where $W(x, t)$ has the form of $W_3(x, t)$ in the proof of Lemma 5.3, but v_1, v_2, v_3 is now a fundamental set of solutions of the differential equation $(\bar{A}_\lambda)$ whose wronskian at a takes the value one. Evidently, with t fixed, $W(x, t)$ is a solution of $(\bar{A}_\lambda)$ having a double zero at t.

If y_1 and $\bar{v}_1$ are respectively solutions of (A_λ) and $(\bar{A}_\lambda)$ satisfying $y_1(a, \lambda) = \bar{v}_1(a, \lambda) = y_1'(a, \lambda) = \bar{v}_1'(a, \lambda) = 0$, $y_1''(a, \lambda) = \bar{v}_1''(a, \lambda) > 0$

and if $x_1 > a$ is the first zero of $\bar{v}_1$ to the right of a, then $W(x, t) \geqq 0$ for $a \leqq t \leqq x \leqq x_1$, by Lemma 1.5. It follows from (5.3) that the first zero of y_1 to the right of a is less than or equal to x_1. This fact and the asymptotic formulae for solutions of ($\bar{A}_\lambda$) (Neumark [111], Chapter II) imply the assertion of Lemma 5.6. □

THEOREM 5.1 (Oscillation Theorem). Let the hypotheses of Lemma 5.5 be met. Let $y(x, \lambda)$ be any non-trivial solution of the differential equation (A_λ) with $y(a, \lambda) = 0$. Given any integer $\nu > 0$, there exists $\lambda_\nu > 0$ such that, for $\lambda > \lambda_\nu$, $y(x, \lambda)$ has at least λ zeros in (a, b).

PROOF. It is sufficient to prove Theorem 5.1 for $y_1(x, \lambda)$. The assertion then follows from properties of bands.

Partition the interval $\langle a, b \rangle$ into $\nu + 1$ equal subintervals by the points

$$x_0 = a < x_1 < \ldots < x_\nu < b = x_{\nu+1} .$$

Assume that for $\lambda = \bar{\lambda} > 0$ every solution $\bar{y}_i(x)$ of (A_λ) satisfying $\bar{y}_i(x_i) = 0$, $\bar{y}_i'(x_i) = 0$, $\bar{y}_i''(x_i) \neq 0$, $i = 1, 2, \ldots, \nu$, has another zero in (x_i, x_{i+1}). Such a $\bar{\lambda}$ exists by Lemma 5.6. It follows from Lemma 5.5 and the properties of bands that it is sufficient to put $\lambda_\nu = \bar{\lambda}$. Thus the theorem is proved. □

LEMMA 5.7. Let the hypotheses of Lemma 5.6 be satisfied. Let $y(x, \lambda)$ be a non-trivial solution of the differential equation (A_λ) with $y(a, \lambda) = 0$. Then the zeros (if they exist) of $y(x, \lambda)$ on the right of a (i.e. in the interval (a, ∞)) are continuous functions of the parameter λ, for $\lambda > 0$.

The proof of Lemma 5.7 is identical with that of Lemma 4.2, and so it is omitted.

THEOREM 5.2. Let the coefficients of the differential equation (A_λ) satisfy the hypotheses of Lemma 5.6. Let $a \leqq b < c < \infty$. Let $\alpha(\lambda)$, $\alpha_1(\lambda)$, $\beta(\lambda)$, $\beta_1(\lambda)$ be continuous functions of the parameter λ and let $|\alpha(\lambda)| + |\alpha_1(\lambda)| \neq 0$, $|\beta(\lambda)| + |\beta_1(\lambda)| \neq 0$ for $\lambda > 0$, while either $\beta(\lambda) \equiv 0$ or $\beta(\lambda) \neq 0$ for all $\lambda > 0$. Then there exist a positive integer ν and a sequence of values of λ (eigenvalues)

$$\lambda_v, \lambda_{v+1}, \ldots, \lambda_{v+p}, \ldots, \quad p=0, 1, \ldots,$$

with a corresponding sequence of functions

$$y_v, y_{v+1}, \ldots, y_{v+p}, \ldots$$

(eigenfunctions) such that $y_{v+p}=y(x, \lambda_{v+p})$ is a solution of (A_λ) satisfying the boundary conditions

$$y(a, \lambda_{v+p})=0,$$

$$\alpha_1(\lambda_{v+p})y(b, \lambda_{v+p})-\alpha(\lambda_{v+p})y'(b, \lambda_{v+p})=0,$$

$$\beta_1(\lambda_{v+p})y(c, \lambda_{v+p})-\beta(\lambda_{v+p})y'(c, \lambda_{v+p})=0$$

and $y(x, \lambda_{v+p})$ has exactly $v+p$ zeros in (a, c).

The proof is similar to that of Theorem 4.7, so it is omitted.

§6. CONDITIONS FOR DISCONJUGATENESS, NON-OSCILLATORICITY AND OSCILLATORICITY OF SOLUTIONS OF THE DIFFERENTIAL EQUATION (A)

In this and the subsequent sections, we shall sometimes denote the differential equations (A), (B), (C), (A_1), (A_2), (B_1), (B_2), using the operator notation, by $L[y]=0$, $M[z]=0$, $F(y, w)=0$, $L_1[y]=0$, $L_2[y]=0$, $M_1[z]=0$, and $M_2[z]=0$, respectively, where L, M, F, L_1, L_2, M_1, M_2 are differential operators on the left-hand sides of the equations (A), (B), (C), (A_1), (A_2), (B_1), (B_2), respectively, applied to functions y, z having continuous, prescribed derivatives in the given interval.

In §6 we treat results by Gera [27, 29—31], some results from his paper [33] and others. In Sections 2, 3, 4 and 5, only a survey of results will be given, without proofs.

1. *Conditions for Disconjugateness of Solutions of the Differential Equation (A)*

Let $a<x_0<\infty$. Denote by I the interval (x_0, ∞).

LEMMA 6.1. Let $g_1(x)$, $g_2(x)$ be continuous functions of $x\in I$ which have no common zero in I. If every non-trivial linear combination of the functions $g_1(x)$, $g_2(x)$ has at most one simple zero in I, then some of their linear combinations has no zero in I.

PROOF. Let $\{x_n\}_{n=1}^{\infty}$ be a sequence of points in I, converging to one of the endpoints of I. Put $\varphi_n(x)=c_n g_1(x)+d_n g_2(x)$, where

$$c_n=\frac{g_2(x_n)}{\sqrt{g_1^2(x_n)+g_2^2(x_n)}}, \quad d_n=-\frac{g_1(x_n)}{\sqrt{g_1^2(x_n)+g_2^2(x_n)}}.$$

Clearly, $\varphi_n(x_n)=0$ and $\varphi_n(x)\neq 0$ for $x\in I$, $x\neq x_n$. Since $c_n^2+d_n^2=1$, the sequences $\{c_n\}_{n=1}^{\infty}$, $\{d_n\}_{n=1}^{\infty}$ are bounded and, therefore contain convergent subsequences. Let $\{c_n^1\}_{n=1}^{\infty}$ and $\{d_n^1\}_{n=1}^{\infty}$ be these subsequences and put $\lim_{n\to\infty} c_n^1=c$, $\lim_{n\to\infty} d_n^1=d$. Then, for every $x\in I$,

$$\lim_{n\to\infty}\varphi_n^1(x)=\lim_{n\to\infty}[c_n^1 g_1(x)+d_n^1 g_2(x)]=$$
$$=cg_1(x)+dg_2(x)=\varphi(x).$$

The sequence $\{\varphi_n^1(x)\}_{n=1}^{\infty}$ evidently converges to φ uniformly on every closed subinterval of I. The function φ is not identically equal to zero in I, because $c^2+d^2=1$. We shall show that $\varphi(x)\neq 0$ for $x\in I$.

Suppose the contrary. Let $\varphi(\bar{x})=0$ for some $\bar{x}\in I$. From the hypotheses of the lemma it follows that $\varphi(x)\neq 0$ for $x\in I$, $x\neq\bar{x}$. To be specific, let $\varphi(x)<0$ for $x<\bar{x}$ and $\varphi(x)>0$ for $x>\bar{x}$, $x\in I$. Let η_1, $\bar{\eta}_1$, η_2, $\bar{\eta}_2$ be such that $\eta_1<\bar{\eta}_1<\bar{x}<\bar{\eta}_2<\eta_2\in I$. Since the sequence $\{x_n\}_{n=1}^{\infty}$ converges to one of the endpoints of I, there exists a number N_1 such that for all $n>N_1$ we have $\varphi_n^1(x)\neq 0$ for $x\in\langle\eta_1,\eta_2\rangle$. Write $\min_{x\in\langle\eta_1,\eta_2\rangle}[-\varphi(x)]=m_1$, $\min_{x\in\langle\eta_1,\eta_2\rangle}\varphi(x)=m_2$, $m=\min\{m_1,m_2\}$. In view of the uniform convergence of a subsequence of $\{\varphi_n\}_{n=1}^{\infty}$ to $\varphi(x)$ on every closed subinterval of I, we deduce the following.

For any positive number, in particular for m as defined above, there exists $N>N_1$ such that for all $x\in\langle\eta_1,\bar{\eta}_1\rangle\cup\langle\bar{\eta}_2,\eta_2\rangle$ we have $|\varphi_n'(x)-\varphi(x)|<m$. Also, $\varphi_n^1(x)\neq 0$ for $x\in\langle\eta_1,\eta_2\rangle$ and $n>N$. Let, say, $\varphi_n^1(x)<$

0 for $x \in \langle \eta_1, \eta_2 \rangle$ and $n > N$. The continuity of $\varphi(x)$ in the interval $\langle \bar{\eta}_2, \eta_2 \rangle$ implies that for some $\xi_2 \in \langle \bar{\eta}_2, \eta_2 \rangle$ we have $\varphi(\xi_2) = m_2$. At ξ_2, for $n > N$, we have $|\varphi_n^1(\xi_2) - \varphi(\xi_2)| = m_2 - \varphi_n^1(\xi_2) > m_2$, contradicting the condition $|\varphi_n^1(\xi_2) - \varphi(\xi_2)| < m \leqq m_2$ for $n > N$. Thus the lemma is proved. □

Let us write

$$l_1[y] \equiv y'' + Py' + Qy = 0\,. \tag{6.1}$$

LEMMA 6.2. A necessary and sufficient condition for the second order differential equation $l_1[y] = 0$ with continuous coefficients $P = P(x)$, $Q = Q(x)$ in an open interval I to be disconjugate in I is the existence of a function u having a continuous second derivative in I such that $u(x) > 0$, $l_1[u(x)] \leqq 0$ for $x \in I$.

PROOF. Necessity. Let the differential equation $l_1(x) = 0$ be non-oscillaroty in I, that is, let any non-trivial linear combination of its fundamental system have at most one simple zero in I. By Lemma 6.1, among the linear combinations of the fundamental system there is one which has no zero in I. Denote that solution by y_0. Evidently, the function $u(x) = y_0(x)$ has the claimed properties.

Sufficiency. Let u be a function having a continuous second derivative in I and such that $u(x) > 0$, $l_1(u) \leqq 0$ for $x \in I$. Put $y(x) = u(x)\,Y(x)$. Assuming that y is a solution of the differential equation $l_1(y) = 0$, implies that the function $Y(x)$ satisfies the differential equation

$$Y''u + (2u' + Pu)\,Y' + l_1(u)\,Y = 0\,, \quad x \in I. \tag{6.2}$$

Since $u^{-1} l_1(u) \leqq 0$ in I, the differential equation (6.2) is non-oscillatory in I (Sansone [124]). The assertion now follows from the relation between solutions of the differential equations $l_1(y) = 0$ and (6.2). The lemma is proved. □

Consider the differential equation

$$m_1[z] \equiv (z' - Pz)' + Qz' + Rz = 0\,, \tag{6.3}$$

where $P = P(x)$, $Q = Q(x)$, $R = R(x)$ are continuous functions of $x \in I$.

REMARK 6.1. Differential equations of the above type are sometimes called quasi-linear, because the unknown function is required only to have its first derivative continuous and (as implied by the equation) also to have the derivative $(z'-Pz)'$ continuous. The situation with the differential equations $M[z]=0$, $M_1[z]=0$, $M_2[z]=0$ and $F[y, w]=0$ is analogous.

The differential equation $m_1[z]=0$ is transformed by the substitution $z(x)=v(x)\exp\left(\int_c^x P(t)\,\mathrm{d}t\right)$, where $c\in I$, into

$$v''+(P+Q)v'+(R+PQ)v=0\,,$$

i.e. a differential equation of the form (6.1). In view of this fact we may restate Lemma 6.2 for the differential equation (6.3).

LEMMA 6.3. *A necessary and sufficient condition for the differential equation (6.3) to be disconjugate in I is the existence of a function $u(x)$ with a continuous first derivative in I such that $m_1[u(x)]$ is a continuous function in I and $u(x)>0$, $m_1[u(x)]\leqq 0$ for $x\in I$.*

REMARK 6.2. If the functions P, Q, R are continuous in $\langle x_0, \infty)$, then (6.1) and (6.3) are disconjugate in $\langle x_0, \infty)$ whenever they are disconjugate in I.

LEMMA 6.4. *The differential equation (A) is disconjugate in $\langle x_0, \infty)$ if and only if there is a solution $\bar{z}$ of the differential equation (B) such that $\bar{z}(x)>0$ for $x\in I$ and such that the differential equation $F(y, \bar{z})=0$ is disconjugate in I.*

The assertion follows immediately from properties of bands.

COROLLARY 6.1. *The differential equation (A) is disconjugate in $\langle x_0, \infty)$ if and only if it is disconjugate in I.*

COROLLARY 6.2. *The differential equation (A) is disconjugate in $\langle x_0, \infty)$ if and only if some solution $\bar{y}$ of (A) and some solution $\bar{z}$ of (B) satisfy $[F(\bar{y}, \bar{z})]_{x=x_0}\leqq 0$, $\bar{y}(x)>0$, $\bar{z}(x)>0$ for $x\in\langle x_0, \infty)$.*

COROLLARY 6.3. The differential equation (A) is disconjugate in $\langle x_0, \infty)$ if and only if the differential equation (B) is also disconjugate.

Proofs of the corollaries follow from Lemma 6.4. They can also be found in Gera's paper [27].

LEMMA 6.5. The differential equation (A) is disconjugate in $\langle x_0, \infty)$ if and only if it has a solution $\bar{y}(x)>0$ for $x\in I$ such that the second order differential equation $F(\bar{y}, z)=0$ is disconjugate in I.

PROOF. a) Assume that (A) is disconjugate in $\langle x_0, \infty)$. By virtue of Corollary 6.3, the differential equation (B) is disconjugate in $\langle x_0, \infty)$ also consequently, the solutions y_1 and z_1 of the differential equations (A) and (B), respectively, with $y_1(x_0)=y_1'(x_0)=0$, $y_1''(x_0)=1$ or $z_1(x_0)=z_1'(x_0)=0$, $[(z_1'-pz_1)']_{x=x_0}=1$ also satisfy $y_1(x)>0$ for $x\in I$ or $z_1(x)>0$ for $x\in I$, respectively. Put $\bar{y}(x)=y_1(x)$ for $x\in\langle x_0, \infty)$. The relationship between adjoint differential equations for $x\in\langle x_0, \infty)$ implies that

$$0=z_1L[\bar{y}]+\bar{y}M[z_1]\equiv\frac{d}{dx}F(\bar{y}, z_1),$$

whence

$$F(\bar{y}, z_1)=\text{const}, \quad x_0\leqq x<\infty .$$

On the other hand, $[F(\bar{y}, z_1)](x_0)=0$, therefore $F(\bar{y}, z_1)=0$ for all $x\in\langle x_0, \infty)$. It now follows from Lemma 6.3 that the equation $F(\bar{y}, z)=0$ is disconjugate in I.

b) Let $\bar{y}$ be a solution of (A) with $\bar{y}(x)>0$ for $x\in I$ and let the differential equation $F(\bar{y}, z)=0$ be disconjugate in I. We are going to show that there exists a solution $\bar{z}$ of (B) with $\bar{z}(x)>0$ for $x\in I$ and such that the second order differential equation $F(y, \bar{z})=0$ is disconjugate in I. The claimed assertion of the theorem will then follow from Lemma 6.4.

So, let the equation $F(\bar{y}, z)=0$ be disconjugate in I. According to Lemma 6.1, there exists a solution of this equation with no zeros in I. Denote it by $\bar{z}(x)$. Assume $\bar{z}(x)>0$ for $x\in I$. Since $\bar{y}(x)>0$, $\bar{z}(x)>0$ and $F(\bar{y}, \bar{z})=0$ for $x\in I$, Lemma 6.2 implies that the differential equation $F(y, \bar{z})=0$ is disconjugate in I. It remains to prove that $\bar{z}(x)$ satisfies (B) in I. It follows from the equation

$$\frac{d}{dx} F(\bar{y}, \bar{z})=0$$

that z has a continuous derivative in I and $M[\bar{z}(x)]$ is a continuous function of $x \in I$. Evidently,

$$\bar{z}L[\bar{y}]+\bar{y}M[\bar{z}] \equiv \frac{d}{dx} F(\bar{y}, \bar{z}) .$$

Since $L[\bar{y}]=0$, $F(\bar{y}, \bar{z})=0$ for $x \in I$, necessarily $M[\bar{z}]=0$. Thus the lemma is proved. □

REMARK 6.3. Lemmas 6.2, 6.3, 6.4 and 6.5 hold also for the interval $I \equiv (a, x_0)$ (Gera [29]).

LEMMA 6.6. Let r_1, r_2 be continuous functions of $x \in (\alpha, \infty)$, $a \leqq \alpha < \infty$, and let the differential equations

$$L_2(y)+r_1 y=0\,, \quad L_2(y)+r_2 y=0$$

be disconjugate in (α, ∞). Moreover, let $r_1(x) \leqq r(x) \leqq r_2(x)$ for $x \in (\alpha, \infty)$. Then the differential equation $L(y)=0$ is also disconjugate in (α, ∞).

The proof of the lemma is in Levin's paper [89] as Corollary 1 to Theorem 2.

THEOREM 6.1. A necessary and sufficient condition for the differential equation (A) to be disconjugate in $\langle x_0, \infty)$ is the existence of functions w and w^* such that $w'''(x)$, $w^{*\prime}(x)$, $M[w^*(x)]$ are continuous functions of $x \in I$ while $w(x)>0$, $w^*(x)>0$, $L[w(x)] \leqq 0$ $(\geqq 0)$ for $x \in I$ and the second order differential equations

$$F(y, w^*)=0\,, \quad F(w, z)=0$$

are disconjugate in I.

PROOF. Sufficiency. Let $w(x)$ and $w^*(x)$ be functions with the required properties. We prove the theorem for the case $L[w(x)] \leqq 0$,

$M[w^*(x)] \leqq 0$ for $x \in I$. (In the case when the inequalities are reversed the proof is analogous.) We have to show that under the stated assumptions the differential equation (A) is disconjugate in $\langle x_0, \infty)$. First we show that the differential equations

$$\bar{L}_1[y] \equiv L_2[y] - \frac{L_2[w]}{w} y = 0 ,$$

$$\bar{L}_2[y] \equiv L_2[y] + \frac{M_2[w^*]}{w^*} y = 0$$

are disconjugate in I.

Let

$$\bar{M}_2[z] \equiv M_2[z] - \frac{M_2[w^*]}{w^*} z = 0$$

be the differential equation adjoint to $\bar{L}_2[y] = 0$. Since $\bar{L}_1[w] = 0$, $\bar{M}_2[w^*] = 0$, for $x \in I$ and the differential equations $F(w, z) = 0$, $F(y, w^*) = 0$ are disconjugate in I, we find by Lemmas 6.5 and 6.4, in view of Remark 6.3, that the differential equations $\bar{L}_1[y] = 0$, $\bar{L}_2[y] = 0$ are disconjugate in I. However, since $L[w(x)] \leqq 0$ and $M[w^*(x)] \leqq 0$ in I, the coefficient $r(x)$ satisfies in I the inequality

$$\frac{M_2[w^*(x)]}{w^*(x)} \leqq r(x) \leqq -\frac{L_2[w(x)]}{w(x)} .$$

The differential equations $\bar{L}_1[y] = 0$ and $\bar{L}_2[y] = 0$ being disconjugate in I, the last inequality implies, in view of Lemma 6.6 and Corollary 6.1, that the differential equation (A) is disconjugate in $\langle x_0, \infty)$.

Necessity follows from Lemmas 6.4 and 6.5. The theorem is proved. □

THEOREM 6.2. A necessary and sufficient condition in order that the differential equation (A) be disconjugate in $\langle x_0, \infty)$ is the existence of functions w and w^* such that $w'''(x)$, $w^{*\prime}(x)$, $M[w^*(x)]$ are continuous functions of $x \in I$ and

$$[F(w, w^*)]_{x=x_0} \leqq 0 , \quad w(x) > 0 , \quad w^*(x) > 0 , \tag{6.4}$$

$$(x - x_0) L[w(x)] \leqq 0 , \quad (x - x_0) M^*[w^*(x)] \leqq 0 \tag{6.5}$$

for $x \in I$.

PROOF. Sufficiency. Suppose that there exist functions w and w^* with the assumed properties. We show that the differential equations $F(w, z)=0$, $F(y, w^*)=0$ are disconjugate in I. Theorem 6.1 implies that the differential equation (A) is disconjugate in I.

The Lagrange identity (Sansone [124])

$$\int_{x_0}^{x} (w^*L[w]+wM[w^*])\,dt=$$
$$=F(w, w^*)-[F(w, w^*)]x=x_0$$

and the properties (6.4), (6.5) show that $F(w, w^*)\leqq 0$ for $x\in I$. It follows by Lemma 6.2 and Lemma 6.3 that the differential equations $F(y, w^*)=0$, $F(w, z)=0$ are disconjugate in I.

Necessity follows from Corollary 6.2. Thus the proof is complete. □

REMARK 6.4. The functions w and w^* may be chosen in a particular way, e. g. to satisfy $w(x_0)=w^*(x_0)=0$. If the coefficients of the differential equation (A) are assumed to have derivatives of certain orders, some special results are obtained (Azbelev and Tsalyuk [5]).

THEOREM 6.3. A necessary and sufficient condition for the differential equation (A) to be disconjugate in $\langle x_0, \infty)$ is the existence of functions w_1, w_2 having continuous third derivatives in I, satisfying

$$w_1(x)>0, \quad w_2(x)>0, \quad L[w_1(x)]\leqq 0, \quad L[w_2(x)]\geqq 0$$

for $x\in I$ and such that the differential equations

$$F(w_1, z)=0, \quad F(w_2, z)=0$$

are disconjugate in I.

PROOF. Let w_1, w_2 be functions with the assumed properties. We shall show that there is a function w_1^* such that $w_1^{*\prime}(x)$ and $M[w_1^*(x)]$ are continuous functions of $x\in I$ while $w_1^*(x)>0$, $M[w_1^*(x)]\leqq 0$ for $x\in I$ and the differential equation $F(y, w_1^*)=0$ is disconjugate in I. Theorem 6.1 then implies that (A) is disconjugate in $\langle x_0, \infty)$.

The differential equation $F(w_2, z)=0$ being disconjugate in I, each of its solutions has at most one simple zero in I. Lemma 6.1 implies the existence of at least one solution of $F(w_2, z)$ with no zeros in I. Denote it by w_1^*. Suppose that $w_1^*(x)>0$ for $x\in I$. Since $w_2(x)>0$, $w_1^*(x)>0$ for $x\in I$ and $F(w_2, w_1^*)=0$ for $x\in I$, the differential equation $F(y, w_1^*)=0$ is disconjugate in I (Lemma 6.2). It remains to prove that the functions w_1^*, $M[w_1^*]$ are continuous in I and that $M[w_1^*(x)]\leqq 0$ for $x\in I$. This assertion follows from the equality

$$\frac{\mathrm{d}}{\mathrm{d}x}F(w_2, w_1^*)=0\,,$$

from the properties of w_2 and w_1^* in I and from the identity

$$w_1^*L[w_2]+w_2M[w_1^*]\equiv\frac{\mathrm{d}}{\mathrm{d}x}F(w_2, w_1^*)\,.$$

Since $F(w_2, w_1^*)=0$, $L[w_2]\geqq 0$ for $x\in I$, we have

$$M[w_1^*]=-\frac{w_1}{w_2}L[w_2]\leqq 0 \quad\text{for}\quad x\in I\,.$$

Necessity follows from Lemma 6.5. Thus the theorem is proved. □

COROLLARY 6.4. Let w_1, w_2 be functions having continuous third derivatives in I and the following properties:

$$w_1(x)>0,\quad w_2(x)>0\,,\quad \begin{vmatrix} w_1, & w_2 \\ w_1', & w_2' \end{vmatrix}>0\,,$$

$$L[w_1(x)]\geqq 0\,,\quad L[w_2(x)]\leqq 0$$

for $x\in I$. Then the differential equation (A) is disconjugate in $\langle x_0, \infty)$.

COROLLARY 6.5. Let there exist a function w having its third derivative continuous in I and satisfying $w(x)>0$, $w'(x)>0$ (<0), $L[w(x)]\leqq 0$ $(\geqq 0)$ in I and let $r(x)\geqq 0$ $(\leqq 0)$ for $x\in\langle x_0, \infty)$. Then the differential equation (A) is disconjugate in $\langle x_0, \infty)$.

REMARK 6.5. Corollaries 6.4 and 6.5 are analysed by Gera [29]. However, they are simple applications of Theorem 6.3. One may choose for w_1, w_2 pairs of elementary functions e^{-x}, e^x or 1, e^x, etc. Special results are thus obtained. Theorem 6.3 can be restated as follows.

TEOREM 6.4. A necessary and sufficient condition for the differential equation (A) to be disconjugate in $\langle x_0, \infty)$ is the existence of functions w_1^*, w_2^* such that $w_1^{*\prime}(x)$, $w_2^{*\prime}(x)$, $M[w_1^*(x)]$, $M[w_2^*(x)]$ are continuous functions of $x \in I$,

$$w_1^*(x)>0, \quad w_2^*(x)>0, \quad M[w_1^*(x)] \leqq 0, \quad M[w_2^*(x)] \geqq 0 \quad \text{for } x \in I$$

and the second order differential equations $F(y, w_1^*)=0$, $F(y, w_2^*)=0$ are disconjugate in I. □

REMARK 6.6. The assertions of Theorems 6.1, 6.2, 6.3 and 6.4 hold true also in the interval $(a, x_0\rangle$ provided that $I \equiv (a, x_0)$.

The following sufficient condition for the differential equation (A) to be disconjugate is an immediate consequence of properties of bands.

THEOREM 6.5. Let $p(x) \leqq 0, q(x) \leqq 0$, $r(x) \geqq 0$ and let $q(x) + (x-x_0)r(x) \leqq 0$ for all $x \in \langle x_0, \infty)$. Then the differential equation (A) is disconjugate in $\langle x_0, \infty)$.

PROOF. In view of the assumption that $q(x) \leqq 0$ for $x \in \langle x_0, \infty)$, the differential equation (A_1) is disconjugate in $\langle x_0, \infty)$, hence no solution y_0 of (A_2) having a double zero at x_1, $x_0 \leqq x_1 < \infty$, has another zero to the left of x_1. The same property is enjoyed by every solution y of the differential equation (A) having a double zero at x_1. This results from the method of variation of constants applied to the differential equation (A) written in the form

$$y''' + py'' + qy' = -ry \, .$$

Then no solution z of the differential equation (B) having a double zero at x_1 has another zero to the right of x_1. Let us show that the hypotheses of the theorem imply also that z cannot have another zero to the left of x_1. Then it follows from the relationship between solutions of the adjoint differential equations (A) and (B) (Remark 5.1) that no solution y with a double zero at x_1 has another zero to the right of x_1. The equation (B), integrated twice from x_1 to x yields

$$z' = pz + \int_{x_1}^{x} [-q(t) + (x-t)r(t)]z(t)\,dt + k(x-x_1)\,.$$

For $x < x_1$, we have

$$z(x) = pz - \int_{x}^{x_1} [-q(t) - (t-x)r(t)]z(t)\,dt - k(x_1 - x)\,.$$

The assumption $q(t) + (t-x_0)r(t) \leqq 0$ implies that

$$-q(t) - (t-x)r(t) \geqq -q(t) - (t-x_0)r(t) \geqq 0\,.$$

Therefore it follows from the preceding relation for $z'(x)$ that $z'(x) \leqq 0$ for $x < x_1$. Thus the theorem is proved. □

2. Solutions without Zeros and Their Relation to Oscillatoricity of Solutions of the Differential Equation (A)

To simplify otherwise complicated expressions, we introduce the following notation.

Let $J \equiv (a, \infty)$ and let I be a subinterval of J. By $S^+(I)$ we shall mean the set of non-negative continuous functions on I. By $S^-(I)$ we shall denote the set of non-positive continuous functions on I, and $S_0^+(I)$ will stand for the set of non-negative continuous functions which are not identically zero on any subinterval of I. Analogously, $S_0^-(I)$ will denote the set of non-positive continuous functions which do not vanish identically on any subinterval of I.

Let $h(x)$ be a continuous function on J. We write $h_+(x) \overset{d}{=} \dfrac{h(x)+|h(x)|}{2}$, $h_-(x) \overset{d}{=} \dfrac{h(x)-|h(x)|}{2}$ for $x \in J$.

Let $v(x)$ be a function having a continuous second derivative in J. Then

$$L_1^+[v] \overset{d}{=} v'' + p(x)v' + q_+(x)v \quad \text{for } x \in J\,.$$

Besides, denote

$$E(x, t) \overset{d}{=} e^{\int_t^x p_-(s)\,ds}, \quad (x, t) \in J \times J\,,$$

$$E^+(x, t) \overset{d}{=} e^{\int_t^x p_+(s)\,ds}, \quad (x, t) \in J \times J\,,$$

$$E^-(x, t) \stackrel{\mathrm{d}}{=} \mathrm{e}^{\int_t^x p_-(s)\,\mathrm{d}s}, \quad (x, t) \in J \times J\,.$$

In the hypotheses of the following theorems, the assumption

$$\int_{x_0}^{\infty} E(x_0, \tau)\,\mathrm{d}\tau = \infty, \quad x_0 \in J \tag{E}$$

will frequently occur.

THEOREM 6.6. Let the differential equation (A_1) be disconjugate in J and let $r(x) \in S_0^+(J)$. Then there is a solution y of the differential equation (A) such that $y(x) \neq 0$, $y(x)y'(x) \leqq 0$ for $x \in J$, $\lim_{x\to\infty} y(x) = k \in (-\infty, \infty)$, while $y'(x)$ vanishes at at most one point $x \in J$.

If, moreover, the condition (E) is satisfied and the differential equation $L_1^+[y] = 0$ is disconjugate in J, then $y'(x) \neq 0$ for all $x \in J$.

THEOREM 6.7. Let $q(x) \in S^-(J)$ and $r(x) \in S_0^+(J)$. Then there exists a solution y_0 of the differential equation (A) satisfying

$$y_0(x)y_0'(x) < 0, \quad y_0(x)y_0''(x) > 0 \text{ for } x \in J$$

and

$$y_0(x)[y_0''(x)E(x, x_0)] \in S_0^-(J), \quad x_0 \in J;$$

moreover

$$\lim_{x\to\infty} y_0(x) = k_0 \in (-\infty, \infty), \quad \lim_{x\to\infty} y_0'(x) = 0,$$

$\lim_{x\to\infty} y_0''(x)E(x, x_0) = k_2 \in (-\infty, \infty)$, while for $x \in J$ one has

$$y_0''(x)E(x, x_0) = k_2 + \int_x^{\infty} q(s)E(s, x_0)y_0'(s)\,\mathrm{d}s + {}$$

$$+ \int_x^{\infty} r(s)E(s, x_0)y_0(s)\,\mathrm{d}s.$$

Assuming also that

a) $p(x) \in S^+(J)$, then $y_0(x)y_0'''(x) \in S_0^-(J)$ and $\lim_{x\to\infty} y''(x) = 0$;

b) condition (E) holds, then $k_2 = 0$;

c) $\int_{x_0}^{\infty} E(x_0, s)\,\mathrm{d}s < \infty$, and $\int_{x_0}^{\infty} \left(\int_x^{\infty} E(x_0, s)\,\mathrm{d}s\right)\mathrm{d}x = \infty$ or $\int_{x_0}^{\infty} q(x) \left(\int_x^{\infty} E(x, s)\,\mathrm{d}s\right)\mathrm{d}x = -\infty$, then $k_2 = 0$.

REMARK 6.7. There exist at most two linearly independent solutions of the differential equation (A) having no zeros in J and possessing the properties listed in Theorem 6.7.

THEOREM 6.8. Assume that $q(x)\in S^-(J)$ and $r(x)\in S_0^+(J)$, or let $r(x)\in S_0^+(J)$ and $q(x)\notin S^-(J)$, while

$$\int_{x_0}^{\infty} E(x_0, s)\, ds=\infty, \quad x_0\in J$$

and let the differential equation $L_1^+[y]=0$ be disconjugate in J. Then a necessary and sufficient condition for the differential equation (A) to be non-oscillatory on J is the existence of a solution y of (A) and a number $x_0\in J$ such that $y(x)y'(x)>0$ for $x\in(x_0, \infty)$.

THEOREM 6.9. Suppose that $r(x)\in S_0^+(J)$, the differential equation (A_1) is disconjugate and the differential equation (A) is oscillatory in J. Then every solution y of (A) with no zeros in J satisfies

$$y(x)y'(x)<0 \text{ for } x\in J, \quad \lim_{x\to\infty} y(x)=k\in(-\infty, \infty)\,.$$

If, at the same time, $q(x)\in S^-(J)$, then we also have

$$y(x)y''(x)>0 \text{ for } x\in J, \quad y(x)[y''(x)E(x, x_0)]'\in S_0^-(J), \; x_0\in J,$$

$$\lim_{x\to\infty} y'(x)=0, \quad \lim_{x\to\infty} y''(x)E(x, x_0)=\bar{k}\in(-\infty, \infty),$$

$$y''(x)E(x, x_0)=\bar{k}+\int_x^{\infty} q(s)E(s, x_0)y'(s)\, ds+$$

$$+\int_x^{\infty} r(s)E(s, x_0)y(s)\, ds, \quad x\in J\,.$$

If, moreover, we have

a) $p(x)\in S^+(J)$, then $y(x)y'''(x)\in S_0^-(J)$ and $\lim_{x\to\infty} y''(x)=0$;

b) $\int_{x_0}^{\infty} E(x_0, s)\, ds=\infty$, then $\bar{k}=0$;

c) $\int_{x_0}^{\infty} E(x_0, s)\, ds<\infty$, and $\int_{x_0}^{\infty}\left[\int_x^{\infty} E(x_0, s)\, ds\right] dx=\infty$ or

$\int_{x_0}^{\infty} q(x)\left[\int_x^{\infty} E(x, s)\, ds\right] dx = -\infty$, then $\bar{k}=0$.

THEOREM 6.10. Let the differential equation (A_1) be disconjugate in J and let $r(x) \in S_0^+(J)$. A necessary and sufficient condition for the differential equation (A) to be oscillatory in J is that every solution y of (A) having no zeros in J should satisfy $y(x)y'(x)<0$ for $x \in J$.

COROLLARY 6.6. a) Let the hypotheses of Theorem 6.10 be satisfied. The differential equation (A) is non-oscillatory in J if and only if there is a non-oscillatory solution y (or a solution without zeros in J) of the differential equation (A) such that $y(x_0)y'(x_0) \geqq 0$ at some $x_0 \in J$.

b) If $q(x) \in S^-(J)$ and $r(x) \in S_0^+(J)$, then the differential equation (A) is non-oscillatory if and only if for some non-trivial, non-oscillatory solution y (or a solution with no zeros in J) and for some $x_0 \in J$ we have $y(x_0)y'(x_0) \geqq 0$ or $y(x_0)y''(x_0) \leqq 0$.

3. *Conditions for the Existence of Oscillatory Solutions of the Differential Equation (A)*

In this paragraph, we shall present (Gera [27, 33]) sufficient conditions for the existence of oscillatory solutions of the differential equation (A) under assumptions guaranteeing the regularity of bands of solutions of (A). The results generalize and extend some results given in papers by Lazer [86] and Hanan [61].

Write

$$J_1 \overset{d}{=} \{x \in J;\, p^2(x) \geqq 4q^2(x)\}, \quad J_0 \overset{d}{=} J - J_1,$$

$$\varphi(x, p, q, r) = r(x) \quad \text{for } x \in J_0$$

and

$$\varphi(x, p, q, r) \overset{d}{=} r(x) + \frac{2}{27} p^3(x) - \frac{1}{3} p(x)q(x) -$$

$$-\frac{2}{27} [p^2(x) - 3q^2(x)]^{3/2} \quad \text{for } x \in J_1.$$

(If $q(x) \in S^-(J)$, then $J_0 = \emptyset$ and $J_1 = J$.)

THEOREM 6.11. Let $p(x)\in S^-(J)$ and let the differential equation $L_1^+[y]=0$ be disconjugate in J. Further, let $r(x)\in S_0^+(J)$ and $\int_{x_0}^{\infty} \varphi(x, p, q, r)E(x, x_0)\,dx=\infty$ for $x_0\in J$. Then the differential equation (A) is oscillatory in J.

REMARK 6.8. If $p(x)\equiv 0$ for $x\in J$, we get the result stated in Hanan's paper [61] as Theorem 1.3. Assuming the existence of derivatives of some coefficients, we obtain further special results (Gera [33]).

THEOREM 6.12. Let $r(x)\in S_0^+(J)$, $q(x)\in S^+(J)$ and assume (E) for $x_0\in J$ if $q(x)\not\equiv 0$ for $x\in J$. Moreover let there exists a continuous function $\mu(x)$ for $x\in J$ such that $\mu(x)>0$ for $x\in(x_0, \infty)$, $x_0\in J$, and

$$\liminf_{x\to\infty} \frac{\int_t^x (x-s)E(x_0, s)\,ds}{\mu(x)\int_t^x E(x_0, s)\,ds} \geqq 1$$

for any $t\in J$. Also, let the differential equation (A_1) be disconjugate in J and let the differential equation

$$v''+p(x)v'+[q(x)+\Theta\mu(x)r(x)]v=0$$

be oscillatory in J for some number $\Theta\in(0, 1)$. Then the differential equation (A) is oscillatory in J.

4. *On Uniqueness of Solutions without Zeros of the Differential Equation (A)*

THEOREM 6.13. Let $q(x)\in S^-(J)$, $r(x)\in S_0^+(J)$, and assume that

$$\int_{x_0}^{\infty} r(x)\left(\int_{x_0}^{x} sE(x, s)\,ds\right)dx=\infty, \quad x_0\in J.$$

Then every solution y of the differential equation (A) with $y(x)y'(x)<0$, $y'(x)y''(x)<0$ in J satisfies $\lim_{x\to\infty} y(x)=\lim_{x\to\infty} y'(x)=0$.

THEOREM 6.14. Let $q(x)\in S^-(J)$, $r(x)\in S_0^+(J)$, and

$$\int_{x_0}^{\infty} sE(x_0, s)\, ds = \infty, \quad x_0 \in J .$$

Then every solution y of the differential equation (A) satisfying $y(x)y'(x) < 0$, $y'(x)y''(x) < 0$ in J has also the property $\lim_{x\to\infty} y'(x) = \lim_{x\to\infty} y''(x)E(x, x_0) = 0$.

THEOREM 6.15. Let x_1 and x_2 be numbers in J. Assume that $r(x) \in S^+(J)$ and that the integrals

$$\int_{x_1}^{\infty} q_+(x) \left(\int_{x_2}^{x} E(x, s)\, ds \right) dx,$$

$$\int_{x_1}^{\infty} r(x) \left(\int_{x_2}^{x} (x - s)E(x, s)\, ds \right) dx$$

are convergent. Then the differential equation (A) is non-oscillatory in J.

THEOREM 6.16. Let $q(x) \in S^-(J)$, $r(x) \in S_0^+(J)$ and let there exist a positive number M with

$$\frac{\int_{x_1}^{x} (x - s)E(x_0, s)\, ds}{\int_{x_1}^{x} sE(x_0, s)\, ds} \leqq M \text{ for } x \geqq x_1 \geqq x_0 > 0 ,$$

where x_0, x_1 are numbers in J. Also, let $\lim_{x\to\infty} y(x) = k \neq 0$ for some solution y of the differential equation (A) with $y(x)y'(x) < 0$, $y(x)y''(x) > 0$ in J. Then the differential equation (A) is non-oscillatory in J.

THEOREM 6.17. Let $r(x) \in S_0^+(J)$, $q(x) \notin S^+(J)$, let $L_1^+[y] = 0$ be disconjugate in J and let the differential equation (A) be non-oscillatory in J. Assume also that

$$\int_{x_0}^{\infty} E(x_0, s)\, ds = \infty, \quad x_0 \in J ,$$

whenever $q(x) \notin S^-(J)$. Then there is at most one solution y_0 (up to linear dependence) of the differential equation (A) with

$$y_0(x)y_0'(x)<0 \text{ for } x\in J, \lim_{x\to\infty} y_0(x)=0 .$$

If such a solution y_0 exists, then every solution y of (A) which is linearly independent of y_0 in J satisfies

$$y(x)y'(x)<0 \text{ for } x\in J \text{ and } \lim_{x\to\infty} y(x)\neq 0 ,$$

or there is a number $x_2\in J$ such that $y(x)y'(x)>0$ for $x>x_2$.

THEOREM 6.18. Let $q(x)\in S^+(J)$, let the assumption (E) hold true, and assume that the differential equation (A_1) is disconjugate in J. Moreover, let $r(x)\in S_0^+(J)$ and assume that the differential equation (A) is non-oscillatory in J. Then there exists exactly one solution of the differential equation (A) (up to linear dependence) with $y_0(x)y_0'(x)<0$ for $x\in J$, and for every solution y of (A) which is linearly independent of y_0 on J there exists $x_2\in J$ such that $y(x)y'(x)>0$, $y'(x)y''(x)>0$ for $x>x_2$.

THEOREM 6.19. Let $p(x)\in S^+(J)$. Let the differential equation $L_1^+[y]=0$ be disconjugate in J and let the condition (E) hold whenever $q(x)\notin S^-(J)$. Also, let $r(x)\in S_0^+(J)$ and let the differential equation (A) be oscillatory in J. Then all non-trivial solutions of (A) are oscillatory in J, except one solution y_0 (up to linear dependence), which satisfies $y_0(x)y_0'(x)<0$ in J.

REMARK 6.9. In Gera's paper [33], further sufficient conditions are derived in order that there may exist exactly one solution of the differential equation (A) without zeros in J. In the hypotheses, however, some coefficients are assumed to have a derivative. Therefore we omit these results here.

5. Some Properties of Solutions of the Differential Equation (A) with $r(x)\leqq 0$

In this paragraph, we present several results concerning the solutions of the differential equation (A), assuming that the equation $L_1[y]$ is disconjugate and that $r(x) \leqq 0$ for $x \in J$. The results (Gera [33]) generalize and supplement those by Švec [137] and Ahmad and Lazer [1].

THEOREM 6.20. Let the differential equation (A_1) be disconjugate in J and let $r(x) \in S_0^-(J)$. Then there exists a solution y of the differential equation (A) satisfying $y(x) \neq 0$, $y(x)y'(x) \geqq 0$ for $x \in J$ and such that $y'(x)$ does not vanish at more than one point $x \in J$.

THEOREM 6.21. Let $x_0 \in J$ and $r(x) \in S^-(\langle x_0, \infty))$. If there exists a function $w(x)$ with its third derivative continuous in $\langle x_0, \infty)$ and such that $w(x) > 0$, $w'(x) < 0$, $L[w(x)] \geqq 0$ for $x > x_0$, then the differential equation (A) is disconjugate in $\langle x_0, \infty)$.

THEOREM 6.22. Let the differential equation (A_1) be disconjugate in J and let $r(x) \in S_0^-(J)$. If the differential equation (A) is oscillatory in J, then for every non-trivial, non-oscillatory solution $y(x)$ of (A) there exists a number $x_0 \in J$ such that $y(x)y'(x) > 0$ for $x > x_0$.

THEOREM 6.23. Let $q(x) \in S^-(J)$ and $r(x) \in S_0^-(J)$. A necessary and sufficient condition for the differential equation (A) to be oscillatory in J is that corresponding to every non-trivial, non-oscillatory solution y of (A) there should exist a number $x_0 \in J$ such that $y(x)y'(x) > 0$, $y'(x)y''(x) > 0$ for $x > x_0$.

THEOREM 6.24. Let $q(x) \in S^-(J)$ and $r(x) \in S_0^-(J)$. Then the differential equation (A) is non-oscillatory in J if and only if for some function $w(x)$ having a continuous third derivative in J and for some $x_0 \in J$ one has either

a) $w(x) > 0$, $w'(x) < 0$, $L[w(x)] \geqq 0$ for $x > x_0$ or

b) $w(x) > 0$, $w''(x) \leqq 0$, $L[w(x)] \geqq 0$ for $x > x_0$.

THEOREM 6.25. Let $q(x) \in S^-(J)$, $r(x) \in S_0^-(J)$, and let the differential equation (A) be oscillatory in J. Then there exist two linearly independent oscillatory solutions y_1, y_2 of the differential equation (A) such that every non-trivial linear combination of them is an oscillatory solution in J and the zeros of any two independent solutions which are linear combinations of y_1 and y_2 separate each other.

THEOREM 6.26. Let $q(x) \in S^-(J)$, $r(x) \in S_0^-(J)$ and let $\int_{x_0}^{\infty} r(x)E(x, x_0)\,dx = -\infty$, $x_0 \in J$. Then the differential equation (A) is non-oscillatory in J if and only if for some solution y of (A) and some $x_0 \in J$ we have $y(x)y'(x) < 0$ for $x > x_0$.

THEOREM 6.27. Let $q(x) \in S^-(J)$, $r(x) \in S_0^-(J)$ and let the differential equation

$$v''' + pv'' + rv = 0$$

be oscillatory in J. Then the differential equation (A) is non-oscillatory if and only if for some solution y of (A) and for some $x_0 \in J$ we have $y(x)y'(x) < 0$ for $x > x_0$.

THEOREM 6.28. Let $p(x) \in S^+(J)$, $q(x) \in S^-(J)$, $r(x) \in S_0^-(J)$ and let the differential equation

$$u''' + qu' + ru = 0$$

or

$$v''' + rv = 0$$

be oscillatory in J. Then the differential equation (A) is non-oscillatory in J if and only if there exist a solution y of (A) and a number $x_0 \in J$ with $y(x)y'(x) < 0$ for $x > x_0$.

§ 7. COMPARISON THEOREMS FOR DIFFERENTIAL EQUATIONS OF TYPE (A) AND THEIR APPLICATIONS

In this section, we give a survey of results concerning comparison of coefficients of differential equations of the type (A) (Gera [33]), their consequences for oscillatoricity of solutions of such differential equations and a remark on asymptotic properties of solutions of the differential equation (A).

1. Comparison Theorems

Along with the differential equation (A), consider the following differential equations

$$l_i[y] \equiv y''' + p_i y'' + q_i y' + r_i y = 0, \quad i = 1, 2,$$

and

$$\bar{l}_1[y] \equiv y''' + p y'' + q_1 y' + r_1 y = 0,$$

where $p_i = p_i(x)$, $q_i = q_i(x)$, $r_i = r_i(x)$, $i = 1, 2$, are continuous functions of $x \in J$.

THEOREM 7.1. Let $r(x) \in S_0^+(J)$ and assume that the differential equation $L_1^+[y] = 0$ is disconjugate in J. Moreover, let the condition (E) be satisfied if $q(x) \notin S^-(J)$. Assume also that $q_1(x) \leqq q(x)$, $0 \leqq r_1(x) \leqq r(x)$ for $x \in J$. If the differential equation $\bar{l}_1[y] = 0$ is oscillatory in J, then so is the differential equation (A) in J.

COROLLARY 7.1. Let $r(x) \in S_0^+(J)$ and let $q_1(x) \leqq q(x) \leqq 0$, $0 \leqq r_1(x) \leqq r(x)$ for $x \in J$. If the differential equation $\bar{l}_1[y] = 0$ is oscillatory in J, then so is the differential equation (A).

Theorem 7.1 can be restated in the following manner.

THEOREM 7.2. Let the hypotheses of Theorem 7.1 be satisfied. If the differential equation (A) is non-oscillatory in J, then the differential equation $\bar{l}_1[y] = 0$ is also non-oscillatory in J.

THEOREM 7.3. Let $r(x) \in S_0^+(J)$, $q(x) \in S^+(J)$ and let (E) hold. Moreover, let the differential equation (A_1) be disconjugate and let

$$p_1(x) \leqq p(x), \; q_1(x) \leqq q(x), \; 0 \leqq r_1(x) \leqq r(x) \quad \text{for } x \in J.$$

If the differential equation $l_1[y] = 0$ is oscillatory in J, then so is the differential equation (A).

COROLLARY 7.2. Let $r(x) \in S_0^+(J)$, $q(x) \in S^+(J)$, $p(x) \in S^+(J)$ and let the assumption (E) be met. Also, suppose that the differential equation

(A_1) is disconjugate in J and that $0 \leqq r_1(x) \leqq r(x)$ for $x \in J$. If the differential equation

$$y''' + r_1 y = 0$$

is oscillatory in J, then the differential equation (A) is oscillatory in J also.

THEOREM 7.4. Let the hypotheses of Theorem 7.3 be satisfied. If the differential equation (A) is non-oscillatory in J, then so is also the differential equation $l_1[y] = 0$.

THEOREM 7.5. Let $r(x) \in S_{\bar{0}}^-(J)$ and let the inequalities $p_2(x) \leqq p(x)$, $q_1(x) \leqq q(x) \leqq q_2(x) \leqq 0$, $r(x) \leqq r_i(x) \leqq 0$, $i = 1, 2$, hold for $x \in J$. Also, let the differential equations $\bar{l}_1[y] = 0$, $l_2[y] = 0$ be oscillatory in J. Then the differential equation (A) is also oscillatory in J.

THEOREM 7.6. Let $q(x) \in S^-(J)$, $r(x) \in S_{\bar{0}}^+(J)$, and let $\int_{x_0}^{\infty} r(x) E(x, x_0)\, dx = -\infty$, $x_0 \in J$. Also, let $q_1(x) \leqq q(x)$, $r(x) \leqq r_1(x) \leqq 0$ for $x \in J$. If the differential equation $\bar{l}_1[y] = 0$ is oscillatory in J, then so is the differential equation (A).

THEOREM 7.7. Let $q(x) \in S^-(J)$, $r(x) \in S_{\bar{0}}^-(J)$ and let the differential equation

$$y''' + p(x)y'' + r(x)y = 0$$

be oscillatory in J. Moreover, let $q_1(x) \leqq q(x)$, $r(x) \leqq r_1(x) \leqq 0$ for $x \in J$. Then the oscillatoricity of the differential equation $\bar{l}_1[y] = 0$ in J implies the oscillatoricity of the differential equation (A) in J.

THEOREM 7.8. Let $p(x) \in S^+(J)$, $q(x) \in S^-(J)$, $r(x) \in S_{\bar{0}}^-(J)$ and let the differential equation

$$y''' + r(x)y = 0$$

be oscillatory in J. Assume also that $q_1(x) \leqq q(x)$, $r(x) \leqq r_1(x) \leqq 0$ for $x \in J$. Then the assertion of Theorem 7.7 holds true.

2. A Simple Application of Comparison Theorems

Assume that the differential equation (A_1) is disconjugate in J and that $r(x) \in S_0^+(J)$. Then the differential equation (C) is the equation of a regular band at $x_0 \in J$ if $w(x_0) = w'(x_0) = 0$, $w''(x_0) \neq 0$, since $w(x) \neq 0$ for $x > x_0$.

In the differential equation (C), consider a special choice of w. Let w be a solution of the first order equation $w' - pw = 0$. Then

$$w = w(x_0) \exp\left(\int_{x_0}^{x} p \, dt\right), \quad a < x_0 < \infty .$$

Let $w(x_0) = 1$. The differential equation (C) takes the form

$$wy'' + qwy = 0 . \tag{C_0}$$

By differentiating (C_0) we obtain

$$wy''' + w'y'' + qwy' + (qw)'y = 0 ,$$

that is,

$$y''' + py'' + qy' + \frac{(qw)'}{w} y = 0 .$$

If w is to be a solution of the differential equation (B), then necessarily $(qw)' = rw$, hence after integration we get the following relation connecting p, q, and r:

$$q_0(x) = e^{-\int_{x_0}^{x} p(t) \, dt} \left[q_0(x_0) + \int_{x_0}^{x} r(t) \, e^{\int_{x_0}^{t} p(s) \, ds} dt \right] , \tag{7.1}$$

where x_0, $x \in J$.

Denote by (A_0) the following equation:

$$y''' + py'' + q_0 y' + ry = 0 . \tag{A_0}$$

LEMMA 7.1. Let $\int_{x_0}^{\infty} r(t) \exp\left(\int_{x_0}^{t} p(s) \, ds\right) dt = k < \infty$ and let $q_0(x_0) + k \leqq 0$. Then the differential equation (A_0) is disconjugate in $\langle x_0, \infty)$.

PROOF. It follows from the assumptions about $q_0(x)$ for $x \geqq x_0$ that the differential equation (C_0) is disconjugate in $\langle x_0, \infty)$ if $q = q_0(x)$. By Lemma 5.3, the bands of solutions of (A_0) are regular to the right of the band point in (x_0, ∞) and owing to the disconjugateness of the differential equation (C_0), in which $q = q_0$, every band of the differential equation (A_0) is disconjugate, i. e. every solution of the differential equation (A_0) has at most two zeros or one double zero in $\langle x_0, \infty)$. Thus the lemma is proved.

Lemma 5.3, Lemma 7.1, Comparison Theorem 7.2 and Corollary 7.1 imply the following theorem.

THEOREM 7.9. Suppose that the coefficients of the differential equation (A) satisfy the hypotheses of Lemma 7.1 and let $q(x) \leqq q_0(x)$ for $x \in \langle x_0, \infty)$. Then the differential equation (A) is non-oscillatory in J.

3. Remark on Asymptotic Properties of Solutions of the Differential Equation (A)

Besides the differential equation (A), consider the following differential equation:

$$z''' + pz'' + qz' + r_1 z = 0\,, \tag{A_z}$$

where $r_1 = r_1(x)$ is a continuous function of $x \in J$. The following theorem holds.

THEOREM 7.10. Let the coefficients of the differential equations (A) and (A_z) satisfy the inequality

$$\int_{x_0}^{\infty} |r_1(x) - r(x)|\, e^{\int_{x_0}^{x} p(s)\,ds} \quad dx < \infty$$

and assume that every solution of (A_z), together with its first derivative, is bounded in $\langle x_0, \infty)$ (for tends to zero as $x \to \infty$). Then every solution of the differential equation (A) has the same property.

The method of proof is similar to that used for Theorems 3.15 and 3.16.

THEOREM 7.11. Let the differential equation (A_1) be disconjugate in J and let $r(x) \in S^{-}(J)$. Also, assume that $r(x) \exp\left(\int_{x_0}^{x} p(t)\, dt\right) < -k^2$ for $x > x_0$ (where $k > 0$ is a constant) and that $f(x) > 0$ is a function with continuous third derivative, satisfying $L[f(x)] \leqq 0$ for $x \geqq x_0$. Then for every solution y of the differential equation (A) without zeros in (x_0, ∞) there exists $x_1 \geqq x_0$ such that

$$|y(x)| - cf(x) > 0 \quad \text{for} \quad x \geqq x_1 ,$$

where c is a suitable positive constant.

Theorem 7.11 gives the rate of divergence of solutions without zeros of the differential equation (A) as $x \to \infty$.

Chapter III

Concluding Remarks

In these concluding remarks we point out a number of results concerning special differential equations of the third order and some questions regarding mutual transformations of solutions of two third order differential equations.

1. Special Forms of Third Order Differential Equations

In 1967, John H. Barrett compiled results on the oscillation theory of ordinary differential equations, which were later published as a separate chapter in the monograph by Mckelvey [96]. An essential part of J. H. Barrett's work deals with the theory of third order linear differential equations. The aim is, first of all, to analyse the so-called typical properties of solutions of third order differential equations and to investigate the methods of proofs of individual assertions. A great deal of his work is devoted to the so-called canonical forms of the third order differential equation (Barrett [7]). The differential equation of the third order is given in the following canonical form:

$$[p_2(p_1y')' + q_1y]' + q_2(p_1y') = 0\,, \tag{A_c}$$

where $p_i > 0$, and the q_i $(i = 1, 2)$ are continuous functions of $x \in (a, b)$. Every differential equation of the form (A) can be reduced to the above canonical form (A_c). To do so, it is sufficient to multiply the equation (A) by $s = \exp(\int p\,\mathrm{d}x)$, which yields

$$(sy'')' + sqy' + sry = 0$$

or

$$[sy''+(\int sr\,dx)y]'+[sq-(\int sr\,dx)]y'=0\,.$$

In view of the canonical form (A_c), the equations

$$(y''+py)'=0 \text{ or } y'''+py'=0$$

are more typical of third order differential equations than is $y'''+p(x)y=0$.

This fact has proved to be of some importance when predicting oscillation properties of solutions of third order differential equations.

The left-hand side of the differential equation (A_c) is the so-called Shinn quasi-differential operator (Shina [133]). The paper by Myshkis and Eventov [98] contains an interesting geometrical approach to oscillation properties of solutions of the third order differential equation in the canonical form

$$y'''+[p(x)y]'+q(x)y'=0\,,$$

where p, q are continuous functions of $x\in(a, b)$.

Special forms of third order differential equations were studied in the following papers: Greguš [51], Greguš and Abdel Karim [57—59], and Vencko [141]. These papers study properties of the solutions of differential equations of the form

$$[py']''+qy=0,\ [py']''+[py']'+qy'+ry=0\,,$$

where $p=p(x)>0$, $q=q(x)$, $r=r(x)$ are continuous functions of $x\in(a, b)$.

Literature on the properties of solutions of the binomial third order differential equation $y'''+p(x)y=0$, where p is a continuous function of $x\in(a, b)$, is very rich. We mention only the papers by Villari [143], Švec [135], Kondratev [79] and others.

The third order differential equation of the form

$$y'''+py''+2Ay'+(A'+b)y=0\,,$$

where $p=p(x)$, $A=A(x)$, $A'=A'(x)$, $b=b(x)$ are continuous functions of $x\in(a, b)$, has been investigated by Moravský [101, 102] and Laitoch [83].

Besides these results, many other authors have treated the third order differential equation (Lasota [85], Regenda [118], Dolan [23], Červeň [20], Kondratev [80], Pfeiffer [113], Azbelev and Tsalyuk [5]). Each of those results extends in some way the theory of the third order differential equation (Hustý [66], Švec [138]).

2. *Remark on Mutual Transformation of Solutions of Third Order Differential Equations*

Šeda [129—131] has constructed the so-called transformation theory for linear differential equations of order n, starting from the results by Borůvka [13—18] on transformations of second order linear differential equations. Moravčík [99] applied these results to differential equations of the third order and derived some interesting results concerning asymptotic and oscillation properties of third order differential equations of the form (a). He also generalized the classical Floquet theory to linear differential equations with non-periodic coefficients.

We present here the basic ideas of this transformation theory.

Consider the following two differential equations of the third order

$$p_0 y''' + p_1 y'' + p_2 y' + p_3 y = p_4 , \tag{1}$$

$$q_0 v''' + q_1 v'' + q_2 v' + q_3 v = q_4 , \tag{2}$$

where $p_i = p_i(x)$, $i = 0, 1, 2, 3, 4$, are continuous functions of x in an interval I_1 and $q_i = q_i(\xi)$, $i = 0, 1, 2, 3, 4$, are continuous functions of ξ in an interval I_2, and $p_0(x) \neq 0$ for $x \in I_1$ and $q_0(\xi) \neq 0$ for $\xi \in I_2$.

The crucial notion in the transformation theory is that of *equivalence* of two differential equations.

We say that the differential equation (2) in the interval I_2 is *equivalent* to the differential equation (1) in I_1 if there exists a pair of function $\xi(x)$, $t(x)$ with the following properties:

1. $\xi(x)$, $t(x)$ are defined in an interval $I_{1\xi} \subset I_1$, $\xi(I_{1\xi}) = I_{2\xi} \subset I_2$,
2. $\xi(x)$, $t(x)$ have continuous third derivatives in $I_{1\xi}$,
3. $\xi'(x) \neq 0$, $t(x) \neq 0$ for $x \in I_{1\xi}$,

and such that:

whenever $v(\xi)$ is a solution of the differential equation (2) in $I_{2\xi}$, then the composition

$$y(x) = t(x)v[\xi(x)] \tag{3}$$

satisfies the differential equation (1) in I_1. The pair of functions $\xi(x)$, $t(x)$ is called a *support* of this equivalence.

It follows from the definition that if we multiply the differential equation (1) by a continuous function $f(x) \neq 0$ for $x \in I_1$, and the differential equation (2), by a continuous function $g(\xi) \neq 0$ for $\xi \in I_2$, we obtain equivalent differential equations, provided that the original equations were equivalent.

The equivalence relation has the following properties.

a) If (2) in $I_{2\xi}$ is equivalent to (1) in $I_{1\xi}$, then (1) in $I_{1\xi}$ is equivalent to (2) in $I_{2\xi}$. Moreover, the solutions $y(x)$ and $v(\xi)$ satisfy for $\xi \in I_{2\xi}$ the inverse relation

$$v(\xi) = \frac{1}{t[x(\xi)]} y[x(\xi)],$$

where $x(\xi)$ is the inverse of $\xi(x)$.

Together with the differential equations (1) and (2), consider the differential equation

$$r_0 w''' + r_1 w'' + r_2 w' + r_3 w = r_4, \tag{4}$$

where $r_i = r_i(\eta)$, $i = 0, 1, 2, 3, 4$, are continuous functions of η in an interval I_3 and $r_0(\eta) \neq 0$ for $\eta \in I_3$.

b) If the differential equation (4) in $I_{3\eta} \subset I_3$ is equivalent to (2) in $I_{2\xi} \subset I_3$ and the differential equation (2) in $I_{2\xi}$ is equivalent to (1) in $I_{1\xi}$, then (4) in $I_{3\eta}$ is equivalent to (1) in $I_{1\xi}$. If $\eta(\xi)$, $u(\xi)$ is a support of the first equivalence and if $\xi(x)$, $t(x)$ is a support of the second equivalence, then $\eta[\xi(x)]$, $t(x)u[\xi(x)]$ is a support of the equivalence of the differential equations (4), (1).

The properties a) and b) imply the following corollary.

The set of linear differential equations can be partitioned into classes of mutually equivalent differential equations. A homogeneous differential equation cannot be equivalent to a non-homogeneous one. If two non-homogeneous differential equations are equivalent, then so are the corresponding homogeneous equations.

c) If the differential equation (2) in $I_{2\xi}$ is equivalent to (1) in $I_{1\xi}$, then there is exactly one support of this equivalence if and only if the pair x, 1 is the only support of the equivalence of (1) in $I_{1\xi}$ to itself and, at the same time, the pair ξ, 1 is the only support of the equivalence of (2) in $I_{2\xi}$ to itself.

In the transformation theory of differential equations, the properties of this partition into classes of equivalent differential equations are studied and the most suitable representatives of the individual classes are sought (such as simple differential equations with constant coefficients).

Moravčík [99] has derived a necessary and sufficient condition for the differential equation (a) to be equivalent to a differential equation with constant coefficients. He has shown that the supports of the equivalence depend substantially on the Laguerre invariant $b = b(x)$.

An important contribution to the transformation theory for linear differential equations of order n is due to Neuman [105—110]. He deals with the so-called global transformation of linear differential equations. The method used is advantageous when solving problems concerning the existence of differential equations whose solutions are to have prescribed properties. It often makes it possible to decide, without going through explicit calculations, whether a certain result is correct or not. In [109] Neuman analyses in this way some results on oscillatory character of solutions of third order differential equations.

In the global transformation theory of linear differential equations, F. Neuman employs results and methods of algebra, geometry, topology, and also methods of the theory of functional equations, enriching these disciplines in various ways.

Chapter IV

Applications of Third Order Linear Differential Equation Theory

This chapter on some applications of the third order linear differential equation theory falls into two parts. In the first part we indicate some applications of the theory to the solution of certain boundary-value problems for non-linear third order differential equations and some possibilities of applying the research methods to certain types of non-linear equations. In the second part we shall deal with the solution of certain problems in physics and engineering which lead to linear third order differential equations. Some of these will be analysed in detail, the rest will only be mentioned, with reference to the appropriate literature.

The main reason for dealing, in the first part, with applications to the theory of non-linear third order equations is that majority of non-linear third order differential equations are equations with a cylindrical phase space and the knowledge of how to solve them is important when attacking the engineering problem of phase synchronization in phase systems of automatic frequency compensation, a topic which has many immediate applications.

§8. SOME APPLICATIONS OF LINEAR THIRD ORDER DIFFERENTIAL EQUATION THEORY TO NON-LINEAR THIRD ORDER PROBLEMS

1. Application of Quasi-linearization to Certain Problems Involving Ordinary Third Order Differential Equations

We are going to apply quasi-linearization (Ghizzetti [37]) to a certain initial and boundary-value problem of the third order, employing some results of the linear third order differential equation theory (Greguš [47]).

Let x be a point in the Euclidean space R_n. Let a non-linear differential equation be given in the form

$$E(u)=f(x, u), \quad x\in A, \tag{8.1}$$

where E is a linear differential operator (ordinary if $n=1$, and partial if $n>1$), $f(x, u)$ is a non-linear function in u, and A is a bounded domain (i.e. a connected open set) in R_n.

We ask whether there exists a solution $u(x)$ of the differential equation (8.1) satisfying the boundary conditions

$$L(u)=0\,, \quad x\in\partial A\,, \tag{8.2}$$

where L is a linear operator and ∂A denotes the boundary of A, i.e. $\bar{A}=A\cup\partial A$.

The problem posed in this way was studied in several papers by R. Bellman and R. Kalaba, and the results achieved in this field were presented at CIME in 1964 by Ghizzetti [37].

In the published lecture (Ghizzetti [37]), the following theorems are proved.

THEOREM A. Let the following assumptions be met.

I. The problem

$$E(v)=g(x), \; x\in A\,; \; L(v)=0, \; x\in\partial A, \tag{8.3}$$

where $g(x)$ is a continuous function of $x\in\bar{A}$, has exactly one continuous solution $v(x)$ for $x\in\bar{A}$, expressible in the form

$$v(x)=\int_{\bar{A}} G(x, s)g(s)\,ds\,, \tag{8.4}$$

$G(x, s)$ being the Green function for the problem (8.3). Also,

$$C=\left\|\int_{\bar{A}} G(x, s)\,ds\right\|, \quad C \text{ being a constant,}$$

where $\|\psi(x)\|$ denotes the norm of ψ defined by

$$\|\psi(x)\|=\max_{x\in\bar{A}}|\psi(x)|\,.$$

II. $f(x, u)$ and its derivatives $f_u(x, u)$, $f_{uu}(x, u)>0$ are continuous functions for $x \in D$ and $|u| \leqq \beta$, $\beta > 0$, where $D \subset R_n$ is a domain with $A \subset D$. Moreover,

$$\max_{x \in D,\ |u| \leqq \beta} \begin{cases} |f(x, u)| = M , \\ |f_u(x, u)| = M_1 , \\ |f_{uu}(x, u)| = M_2 , \end{cases}$$

where M, M_1, M_2 are real numbers.

III. The following inequality holds:

$$C(M + 2\beta M_1) \leqq \beta .$$

IV. Let $z(x)$ be a function defined and continuous for $x \in A$, and let $|z(x)| \leqq \beta$ for $x \in A$. Assume that if a function $\varphi(x)$ satisfies

$$E(\varphi) \geqq f_u[x, z(x)]\varphi(x), \quad x \in A,$$

$$L(\varphi) = 0, \quad x \in \partial A,$$

then $\varphi(x) \geqq 0$ ($\varphi(x) \leqq 0$) for $x \in \bar{A}$; also, let $\varphi(x) \equiv 0$ if and only if $E(\varphi) = f_u[x, z(x)]\varphi(x)$, $x \in A$.

Then, on the above assumptions I to IV, there exists exactly one solution $u(x)$ of the problem (8.1), (8.2), given by

$$u(x) = \max_{z(x)} w[x, z(x)]$$

$$(u(x) = \min_{z(x)} w[x, z(x)] \quad \text{in case} \quad \varphi(x) \leqq 0),$$

where $w[x, z(x)]$ is a solution of the linear problem

$$E(w) = f(x, z) + (w - z)f_u(x, z), \quad x \in A;$$

$$L(w) = 0, \quad x \in \partial A \tag{8.5}$$

with a parametric function $z(x)$.

THEOREM B. Let the hypotheses of Theorem A be satisfied, let $z(x)$ be a fixed parametric function, and let $\{w_n(x)\}$ be the sequence defined by

$$E(w_1) = f(x, z) + (w_1 - z)f_u(x, z), \quad x \in A;$$

$$L(w_1) = 0, \quad x \in \partial A, \tag{8.6}$$

$$E(w_{n+1}) = f(x, w_n) + (w_{n+1} - w_n) f_u(x, w_n), \ x \in A;$$

$$L(w_{n+1}) = 0, \ x \in \partial A. \tag{8.7}$$

Then

a) the sequence $\{w_n(x)\}$ is non-decreasing (non-increasing in case $\varphi(x) \leqq 0$),

b) the sequence $\{w_n(x)\}$ converges uniformly to a solution $u(x)$ of the problem (8.1), (8.2).

Using Theorems A and B and results on the third order linear differential equation, we shall prove the following two theorems.

THEOREM 8.1. Let $f(x, u)$, $f_u(x, u) \geqq 0$, $f_{uu}(x, u) > 0$ be continuous functions of $x \in \langle 0, \alpha \rangle$ and $|u| \leqq \beta$. Then there is exactly one solution u of the initial value problem

$$u''' = f(x, u), \ u(0) = u'(0) = u''(0) = 0, \ 0 \leqq x \leqq a,$$
$$0 \leqq a < \alpha \tag{8.8}$$

satisfying the assertions of Theorems A and B.

PROOF. To prove Theorem 8.1, we verify the assumptions I, II, III, and IV of Theorem A.

I. Let $g(x)$ be a continuous function for $0 \leqq x < \alpha$. By variation of constants it is easy to show that the problem

$$v''' = g(x), \ v(0) = v'(0) = v''(0) = 0$$

for $x > 0$ is solved by

$$v(x) = \int_0^x \frac{W(x, t)}{W(t)} g(t) \, dt,$$

where

$$W(t) = \begin{vmatrix} t^2, & t, & 1 \\ 2t, & 2, & 0 \\ 2, & 0, & 0 \end{vmatrix} = -2, \quad W(x, t) = \begin{vmatrix} x^2, & x, & 1 \\ t^2, & t, & 1 \\ 2t, & 1, & 0 \end{vmatrix} = -(x-t)^2$$

and x^2, x, 1 is a fundamental system of solutions of the equation $v''' = 0$. We have thus established that

$$v(x)=\frac{1}{2}\int_0^x (x-t)^2 g(t)\,dt\,.$$

Let $0<a\leqq\alpha$. Clearly,

$$C=\left\|\frac{1}{2}\int_0^a (x-t)^2\,dt\right\|=\max_{0\leqq x\leqq a}\frac{1}{2}\int_0^a (x-t)^2\,dt=$$

$$=\max_{0\leqq x\leqq a}\frac{1}{2}\left[-\frac{(x-t)^3}{3}\right]_0^a=\max_{0\leqq x\leqq a}\frac{1}{2}\left(-\frac{(x-a)^3}{3}+\frac{x^3}{3}\right)=\frac{a^3}{6}\,.$$

II. This assumption is obviously fulfilled.

III. We obtain the following estimate for the length of $\langle 0, a\rangle$.

$$\frac{a^3}{6}(M+2\beta M_1)\leqq\beta\,,$$

i.e.

$$a\leqq\sqrt[3]{\frac{6\beta}{M+2\beta M_1}}\,. \tag{8.9}$$

IV. Let $z(x)$ be a continuous function of $x\in<0, a>$ and let $|z(x)|\leqq\beta$.

We have to show that there exists a solution $\varphi(x)\geqq 0$ for $0\leqq x\leqq a$ to the problem

$$\varphi'''(x)\geqq f_u[x, z(x)]\varphi(x),\quad \varphi(0)=\varphi'(0)=\varphi''(0)=0\,.$$

The problem can be written in the form

$$\varphi'''(x)-f_u[x, z(x)]\varphi(x)=h(x)\geqq 0,$$

$$\varphi(0)=\varphi'(0)=\varphi''(0)=0\,, \tag{8.10}$$

where $h(x)$ is a continuous function of $x\in<0, a>$.

The problem (8.10) is solved by

$$\varphi(x)=\int_0^x W(x, t)h(t)\,dt\,, \tag{8.11}$$

where

$$W(x, t)=\begin{vmatrix}\varphi_1(x), & \varphi_2(x), & \varphi_3(x)\\ \varphi_1(t), & \varphi_2(t), & \varphi_3(t)\\ \varphi_1'(t), & \varphi_2'(t), & \varphi_3'(t)\end{vmatrix}$$

and $\varphi_1, \varphi_2, \varphi_3$ is a fundamental system of solutions of the equation $\varphi''' - f_u(x, z)\varphi = 0$ whose wronskian $W(\varphi_1, \varphi_2, \varphi_3) = 1$.

The formula (8.11) can be obtained again by the variation of constants and it evidently implies that $\varphi(x) \equiv 0$ whenever $h(t) \equiv 0$ for $0 \leqq x \leqq a$. Conversely, $\varphi(x) \leqq 0$ implies $h(x) \leqq 0$. However, for a fixed t, $W(x, t)$ is a solution of a homogeneous third order linear differential equation, having a double zero at t. Applying the theory of the third order equation, we have in this case $W(x, t) > 0$ for $t < x$, which together with (8.11) implies that $\varphi(x) \geqq 0$ for $x \geqq 0$. Thus the theorem is proved. □

COROLLARY 8.1. The right-hand side in (8.9) is an increasing function of β. Therefore,

$$\max a \leqq \lim_{\beta \to \infty} \sqrt[3]{\frac{6\beta}{M + 2\beta M_1}} = \sqrt[3]{\frac{3}{M_1}}\,.$$

COROLLARY 8.2. Let $f(x, 0) > 0$. Then the solution of (8.8) has no other zero in the interval $\langle 0, a)$.

The proof follows from Theorem B by putting $z(x) \equiv 0$ in (8.6).

LEMMA 8.1. Any solution y of the differential equation $y''' + M_1 y = 0$ with a double zero at 0 has no other zero in the interval

$$0 \leqq x \leqq a, \quad a = 3\sqrt[3]{\frac{3}{4}\frac{1}{M_1}}\,.$$

PROOF. Any solution y of the differential equation $y''' + M_1 y = 0$ having a double zero at 0, is of the form

$$y = \frac{1}{3}k^2\left[\mathrm{e}^{-kx} - \mathrm{e}^{kx/2}\cos\frac{\sqrt{3}}{2}kx + \sqrt{3}\,\mathrm{e}^{kx/2}\sin\frac{\sqrt{3}}{2}kx\right],$$

where $k = \sqrt[3]{M_1}$.

Evidently,

$$y=\frac{1}{3}k^2 e^{kx/2}\cdot$$

$$\cdot\left[e^{-(3/2)kx}-2\left(\frac{1}{2}\cos\frac{\sqrt{3}}{2}kx-\frac{\sqrt{3}}{2}\sin\frac{\sqrt{3}}{2}kx\right)\right],$$

$$y=\frac{1}{3}k^2 e^{kx/2}\cdot$$

$$\cdot\left[e^{-3kx/2}+2\left(-\sin\frac{\pi}{6}\cos\frac{\sqrt{3}}{2}kx+\cos\frac{\pi}{6}\sin\frac{\sqrt{3}}{2}kx\right)\right],$$

$$y=\frac{1}{3}k^2 e^{kx/2}\left[e^{-3kx/2}+2\sin\left(\frac{\sqrt{3}}{2}kx-\frac{\pi}{6}\right)\right].$$

The first zero of y to the right of 0 is obviously greater than the second zero x_2 of the function

$$\sin\left(\frac{\sqrt{3}}{2}kx-\frac{\pi}{6}\right).$$

From the fact that

$$\sin\left(\frac{\sqrt{3}}{2}kx-\frac{\pi}{6}\right)=0$$

at x_2 we get

$$\frac{\sqrt{3}}{2}kx_2-\frac{\pi}{6}=\pi,$$

i.e.

$$x_2=\frac{7}{3\sqrt{3}}\frac{\pi}{k}.$$

We now show that $a<x_2$.

Indeed,

$$a=\frac{3}{\sqrt[3]{4}}\sqrt[3]{3}\frac{1}{k}<x_2=\frac{7}{3\sqrt{3}}\frac{\pi}{k}.$$

The lemma is proved. □

THEOREM 8.2. Let the hypotheses of Theorem 8.1 be met. Then there exists exactly one non-trivial solution of the problem

$$u''' = f(x, u), \quad u(0) = u'(0) = u(a) = 0, \quad 0 < a \leqq \alpha \tag{8.12}$$

satisfying the assertions of Theorems A and B.

PROOF. Theorem 8.2 will be established as soon as we show that the hypotheses I to IV of Theorem A are satisfied.

I. Let $g(x)$ be a continuous function for $0 \leqq x \leqq a$. The problem

$$v''' = g(x), \quad v(0) = v'(0) = v(a) = 0 \tag{8.13}$$

has exactly one solution

$$v(x) = \int_0^a G(x, \xi) g(\xi)\, d\xi ,$$

where $G(x, \xi)$ is the Green function for the homogeneous problem corresponding to (8.13).

In fact, for the fundamental system of solutions x^2, x, 1 of the equation $v''' = 0$ with wronskian $W = -2$ we have

$$G(x, \xi) = \begin{cases} a_1 x^2 + a_2 x + a_3 , & 0 \leqq x \leqq \xi , \\ b_1 x^2 + b_2 x + b_3 , & 0 \leqq \xi \leqq x \leqq a . \end{cases}$$

At ξ, it is

$$\begin{aligned} c_1 \xi^2 + c_2 \xi + c_3 &= 0 \\ 2c_1 \xi + c_2 &= 0 \\ 2c_1 &= 1, \quad c_i = b_i - a_i, \quad i = 1, 2, 3, \end{aligned}$$

whence

$$c_1 = \frac{1}{2}, \quad c_2 = -\xi , \quad c_3 = \frac{1}{2} \xi^2 .$$

Using the fact that $G(x, \xi)$ necessarily satisfies the boundary conditions, we deduce that

$$\begin{aligned} &a_3 = 0 , \\ &a_2 = 0 , \\ &b_1 a^2 + b_2 a + b_3 = 0 , \end{aligned}$$

hence

$$a_1 = \frac{\xi}{a} - \frac{\xi^2}{2a} - \frac{1}{2}, \quad b_1 = -\frac{\xi^2}{2a^2} + \frac{\xi}{a}, \quad b_2 = -\xi, \quad b_3 = \frac{1}{2}\xi^2.$$

Thus we obtain the Green function in the form

$$G(x, \xi) = \begin{cases} -\frac{1}{2}\left(1 - \frac{\xi}{a}\right)^2 x^2, & 0 \leqq x \leqq \xi \leqq a, \\ \frac{\xi}{a}\left(1 - \frac{\xi}{2a}\right) x^2 - \xi x + \frac{1}{2}\xi^2, & 0 \leqq \xi \leqq x \leqq a. \end{cases}$$

Further, we have to calculate $C = \left\| \int_0^a |G(x, \xi)| \, d\xi \right\|$.

It is easy to see that at a fixed x for $0 \leqq \xi \leqq a$ we have $G(x, \xi) \leqq 0$. Hence it follows that

$$\int_0^a |G(x, \xi)| \, d\xi = -\int_0^a G(x, \xi) \, d\xi = \frac{x^2}{6}(a - x),$$

$$\max_{0 \leqq x \leqq a} \frac{x^2}{6}(a - x) = \frac{2}{81} a^3 = C.$$

Hypothesis II is obviously satisfied.

III. For the length of the interval $\langle 0, a \rangle$ we obtain the following estimate:

$$\frac{2}{81} a^3 (M + 2\beta M_1) \leqq \beta,$$

hence

$$a \leqq 3 \sqrt[3]{\frac{3}{2} \frac{\beta}{M + 2\beta M_1}}. \tag{8.14}$$

IV. Let $z(x)$ be a continuous function of $0 \leqq x \leqq a$ and let $|z(x)| \leqq \beta$. Our aim is to prove the existence of a solution $\varphi(x) \leqq 0$, $0 \leqq x \leqq a$, to the problem

$$\varphi''' \geqq f_u(x, z)\varphi, \quad \varphi(0) = \varphi'(0) = \varphi(a) = 0.$$

The problem can be rewritten in the form

$$\varphi''' - f_u(x, z)\varphi = h(x) \geqq 0, \quad \varphi(0) = \varphi'(0) = \varphi(a) = 0, \tag{8.15}$$

where $h(x)$ is a continuous function of $0 \leqq x \leqq a$.

Now construct the Green function for the homogeneous problem corresponding to (8.15).

Let φ_1, φ_2, φ_3, be a fundamental set of solutions of the equation $\varphi''' - f_u(x, z)\varphi = 0$ with $\varphi_1(0) = \varphi_1'(0) = 0$, $\varphi_1''(0) = 1$, $\varphi_2(0) = \varphi_2''(0) = 0$, $\varphi_2'(0) = 1$, $\varphi_3'(0) = \varphi_3''(0) = 0$, $\varphi_3(0) = 1$.

Evidently, $W(\varphi_1, \varphi_2, \varphi_3) = -1$.

Then for $G(x, \xi)$ we deduce

$$G(x, \xi) = \begin{cases} a_1\varphi_1(x) + a_2\varphi_2(x) + a_3\varphi_3(x), & 0 \leqq x \leqq \xi \leqq a, \\ b_1\varphi_1(x) + b_2\varphi_2(x) + b_3\varphi_3(x), & 0 \leqq \xi \leqq x \leqq a. \end{cases}$$

At ξ, we obviously have

$$\begin{aligned} &c_1\varphi_1(\xi) + c_2\varphi_2(\xi) + c_3\varphi_3(\xi) = 0, \\ &c_1\varphi_1'(\xi) + c_2\varphi_2'(\xi) + c_3\varphi_3'(\xi) = 0, \\ &c_1\varphi_1''(\xi) + c_2\varphi_2''(\xi) + c_3\varphi_3''(\xi) = 1, \quad c_i = b_i - a_i, \quad i = 1, 2, 3. \end{aligned}$$

Hence it follows that

$$c_1 = -\begin{vmatrix} \varphi_2(\xi), & \varphi_3(\varphi) \\ \varphi_2'(\xi), & \varphi_3'(\xi) \end{vmatrix}, \quad c_2 = \begin{vmatrix} \varphi_1(\xi), & \varphi_3(\xi) \\ \varphi_1'(\xi), & \varphi_3'(\xi) \end{vmatrix},$$

$$c_3 = -\begin{vmatrix} \varphi_1(\xi), & \varphi_2(\xi) \\ \varphi_1'(\xi), & \varphi_2'(\xi) \end{vmatrix}.$$

Since $G(x, \xi)$ is to satisfy the boundary conditions, we get

$$a_2 = 0, \quad a_3 = 0, \quad b_2 = \begin{vmatrix} \varphi_1(\xi), & \varphi_3(\xi) \\ \varphi_1'(\xi), & \varphi_3'(\xi) \end{vmatrix},$$

$$b_3 = -\begin{vmatrix} \varphi_1(\xi), & \varphi_2(\xi) \\ \varphi_1'(\xi), & \varphi_2'(\xi) \end{vmatrix},$$

$$b_1 = \frac{1}{\varphi_1(a)}\left[\varphi_3(a)\begin{vmatrix} \varphi_1(\xi), & \varphi_2(\xi) \\ \varphi_1'(\xi), & \varphi_2'(\xi) \end{vmatrix} - \right.$$

$$\left. - \varphi_2(a)\begin{vmatrix} \varphi_1(\xi), & \varphi_3(\xi) \\ \varphi_1'(\xi), & \varphi_3'(\xi) \end{vmatrix}\right],$$

$$a_1 = \frac{1}{\varphi_1(a)}\left[\varphi_3(a)\begin{vmatrix} \varphi_1(\xi), & \varphi_2(\xi) \\ \varphi_1'(\xi), & \varphi_2'(\xi) \end{vmatrix} - \varphi_2(a)\begin{vmatrix} \varphi_1(\xi), & \varphi_3(\xi) \\ \varphi_1'(\xi), & \varphi_3'(\xi) \end{vmatrix} + \right.$$

$$\left. + \varphi_1(a)\begin{vmatrix} \varphi_2(\xi), & \varphi_3(\xi) \\ \varphi_2'(\xi), & \varphi_3'(\xi) \end{vmatrix}\right] = \frac{1}{\varphi_1(a)} W(a, \xi).$$

Thus we have obtained

$$G(x, \xi) = \begin{cases} \dfrac{1}{\varphi_1(a)} W(a, \xi)\varphi_1(x), & 0 \leqq x \leqq \xi \leqq a, \\ \dfrac{1}{\varphi_1(a)} W(a, \xi)\varphi_1(x) - W(x, \xi), & 0 \leqq \xi \leqq x \leqq a. \end{cases}$$

Evidently, for a fixed ξ, $W(x, \xi)$ is a solution of the differential equation $\varphi''' - f_u(x, z)\varphi = 0$ with a double zero at ξ and $W''_{xx}(\xi, \xi) < 0$.

The differential equation adjoint to $\varphi''' - f_u(x, z)\varphi = 0$ is the equation $\psi''' + f_u(x, z)\psi = 0$. Let $\psi(x)$ be a solution of the latter equation with $\psi(0) = \psi'(0) = 0$, $\psi''(0) < 0$. Comparing the equation $\psi''' + f_u(x, z)\psi = 0$ with $\bar{\psi} + M_1\bar{\psi} = 0$, we conclude by Lemma 8.1 and Theorem 2.5 that the solution ψ has no other zero in $\langle 0, a\rangle$. Then, applying Section 4 of §1 to the adjoint equation $\varphi''' - f_u(x, z)\varphi = 0$, we see that its solution $W(x, \xi)$ for $\xi = 0$ has no other zero in $\langle 0, a\rangle$. From the properties of bands it follows that the solution $W(x, \xi)$, for any fixed $\xi \in \langle 0, a)$, has no other zero to the right of ξ in $\langle 0, a\rangle$.

The integral identity (1.8), applied to the equation $\varphi''' - f_u(x, z)\varphi = 0$ and its solution $W(x, \xi)$ (with ξ fixed), implies that $W(x, \xi)$, has no other zero to the left of ξ. The above reasoning yields the conclusion that $W(x, \xi) \leqq 0$ for $0 \leqq x \leqq a$, $0 \leqq \xi \leqq a$.

We now have to show that $G(x, \xi) \leqq 0$ for any fixed x and every $0 \leqq \xi \leqq a$. Obviously, $G(x, \xi) \leqq 0$ for $a \geqq \xi \geqq x \geqq 0$. We show that $G(x, \xi) \leqq 0$ for $0 \leqq \xi \leqq x \leqq a$ also. With ξ fixed, $G(x, \xi)$ is evidently a solution of $\varphi''' - f_u(x, z)\varphi = 0$ in the interval $0 \leqq x \leqq a$, and $G(a, \xi) = 0$, $G(0, \xi) > 0$, $G(\xi, \xi) > 0$.

It follows, in view of the properties of bands, that the function $G(x, \xi)$ has no other zero in the interval $0 < x < \xi$. Thus $G(x, \xi) \leqq 0$ for $0 \leqq \xi \leqq x \leqq a$.

The solution to the problem (8.15) may be written in the form

$$\varphi(x) = \int_0^a G(x, \xi)h(\xi)\, d\xi$$

and hence clearly $\varphi(x) \leqq 0$ for $0 \leqq x \leqq a$. Thus the theorem is proved. □

Note that by (8.14) the right-hand side is an increasing function of β and hence

$$\max_{\beta} \varphi \leqq \lim_{\beta\to\infty} 3\sqrt[3]{\frac{3}{2}\frac{\beta}{M+2\beta M_1}} = 3\sqrt[3]{\frac{3}{4}\frac{1}{M_1}},$$

where $M_1 = \max_{0\leqq x\leqq a} |f_u(x, z(x))|$.

2. Three-point Boundary-value Problems for Third Order Non-linear Ordinary Differential Equations

These problems have been studied by several authors; these include Baar and Sherman [6], Bespalova and Klokov [11], Jackson [67], Jackson and Schrader [68], Klassen [78], Schrader [128], Šeda [132], Kiguradze [77] to mention only a few. The topic was summarized and supplemented by original results by I. Rusnák in his thesis, where he used the results of §4 of this monograph. Here we give an application of the Green function corresponding to a linear three-point boundary-value problem of the third order.

We shall deal with a boundary-value problem of the type

$$y''' = f(x, y, y', y''), \quad (x, y, y', y'') \in \langle a_1, a_3\rangle \times R^3, \tag{8.16}$$

$$\begin{aligned} \alpha_1 y(a_1) + \alpha_2 y'(a_1) + \alpha_3 y''(a_1) &= A_1, \\ \beta_1 y(a_2) + \beta_2 y'(a_2) + \beta_3 y''(a_2) &= A_2, \\ \gamma_1 y(a_3) + \gamma_2 y'(a_3) + \gamma_3 y''(a_3) &= A_3, \end{aligned} \tag{8.17}$$

where $a_i, \alpha_i, \beta_i, \gamma_i, A_i \in R$, $i = 1, 2, 3$, $a_1 < a_2 < a_3$,

$$\sum_{i=1}^{3} |\alpha_i| > 0, \quad \sum_{i=1}^{3} |\beta_i| > 0, \quad \sum_{i=1}^{3} |\gamma_i| > 0.$$

An analogous problem for a second order equation was studied by Schmidt [127].

Applying Theorems 4.1 and 4.2, it can be shown that any solution y of the boundary-value problem (8.16), (8.17) solves the integro-differential equation

$$y(x)=\varphi(x)+\sum_{k=1}^{2}\int_{a_k}^{a_{k+1}} G_k(x, s)f(s, y(s), y'(s), y''(s))\, ds \tag{8.18}$$

and conversely, $G_k(x, s)$ being particular Green's functions for the boundary-value problem given by $y'''=0$ and the homogeneous boundary conditions obtained from (8.17) by putting $A_1=A_2=A_3=0$, making the assumption, of course, that there exists only the trivial solution to that problem. A necessary and sufficient condition for this is that

$$\Delta=\begin{vmatrix} \alpha_1, & \alpha_1 a_1+\alpha_2, & \alpha_1 a_1^2+2\alpha_2 a_1+2\alpha_3 \\ \beta_1, & \beta_1 a_2+\beta_2, & \beta_1 a_2^2+2\beta_2 a_2+2\beta_3 \\ \gamma_1, & \gamma_1 a_3+\gamma_2, & \gamma_1 a_3^2+2\gamma_2 a_3+2\gamma_3 \end{vmatrix}\neq 0 .$$

We assume in the sequel that $\Delta\neq 0$. The function $\varphi(x)$ in (8.18) is a solution of the boundary-value problem given by $y'''=0$ and the boundary conditions (8.17).

Using the Schauder theorem (which we do not state here) (see Hartman [62]), we shall prove an existence theorem for the boundary-value problem (8.16), (8.17).

THEOREM 8.3. Let f be continuous on $\langle a_1, a_3\rangle\times R^3=I\times R^3$ and let there exist a constant $M>0$ with

$$|f(x, y, y', y'')|\leqq M \quad \text{for all } (x, y, y', y'')\in I\times R^3 .$$

Then the boundary-value problem (8.16), (8.17) has at least one solution.

PROOF. Let $C_2(I)$ be a Banach space equipped with the norm

$$\|y\|=\sum_{i=1}^{2}\max_I |y^{(i)}(x)| .$$

Define an operator T: $C_2\to C_2$ by

$$Ty(x)=\varphi(x)+\sum_{k=1}^{2}\int_{a_k}^{a_{k+1}} G_k(x, s)f(s, y(s)y'(s), y''(s))\, ds .$$

Further, introduce the constants

$$K=\max_I|\varphi(x)|,\quad K'=\max_I|\varphi'(x)|,\quad K''=\max_I|\varphi''(x)|,$$

$$N=(a_3-a_1)\max\{\sup_{I\times\langle a_1,a_2\rangle}|G_1(x,s)|,\ \sup_{I\times\langle a_1,a_2\rangle}|G_2(x,s)|\},$$

$$N'=(a_3-a_1)\max\{\sup_{I\times\langle a_1,a_2\rangle}|G'_{1x}(x,s)|,\ \sup_{I\times\langle a_1,a_2\rangle}|G'_{2x}(x,s)|\},$$

$$N''=(a_3-a_1)\max\{\sup_{I\times\langle a_1,a_2\rangle}|G''_{1xx}(x,s)|,\ \sup_{I\times\langle a_1,a_2\rangle}|G''_{2xx}(x,s)|\}.$$

Then we have

$$|Ty(x)|\leqq K+MN,\quad |(Ty)'(x)|\leqq K'+MN',$$
$$|(Ty)'(x)|\leqq K''+MN''\ \text{for all}\ x\in I.$$

Now define a set E by

$$E=\{y\in C_2(I): |y|\leqq K+MN,\ |y'|\leqq K'+MN',$$
$$|y''|\leqq K''+MN''\}.$$

E is a closed and convex subset of $C_2(I)$, $TE\subset E$, TE is relatively compact and T is a continuous operator. Thus, the hypotheses of Schauder's fixed point theorem concerning T are fulfilled and therefore there exists at least one fixed point $y(x)\in E$ of the operator T satisfying (8.18). This fixed point is a solution of the boundary-value problem (8.16), (8.17). Thus the theorem is proved. □

3. *On Properties of Solutions of a Certain Non-linear Third Order Differential Equation*

We are going to examine briefly a non-linear third order differential equation of the form

$$y'''+p(x)y'+q(x)y^r=0, \tag{8.19}$$

where $p(x)$, $q(x)$ are continuous functions defined on $\langle a,\infty)$, $a>0$. Moreover, we assume that $q(x)$ is not identically zero for large x and that the exponent r is a quotient of odd positive integer. This guarantees that the solutions with real initial conditions be real and that the negative of any solution of (8.19) is again a solution of (8.19).

The equation (8.19) was extensively studied by Heidel [64], and we shall present the essence of his results. The study of this equation was motivated by the results and methods of research on solutions of the linear third order differential equation and also by the study of the properties of solutions of the equation (8.19) (Waltman [144]), and by the research on similar types of equations (Kiguradze [75, 76], Ličko and Švec [90]). Special mention must be made of the monograph (Reissig et al. [119]).

A solution of the equation (8.19) is said to be extendable if it exists on $\langle a_1, \infty)$ for some $a_1 \geqq a$.

A non-trivial solution of (8.19) is called oscillatory if it is extendable and has zeros at arbitrarily large x.

A non-trivial solution of the equation (8.19) is said to be non-oscillatory if it is extendable and not oscillatory.

We shall be primarily interested in extendable solutions, although we shall show in the first two theorems that in case $r \leqq 1$ all the solutions of (8.19) are extendable and that under certain assumptions the non-extendable solutions of (8.19) have infinitely many zeros in a finite interval.

THEOREM 8.4. If $r \leqq 1$ and $\langle x_0, b\rangle$ is any compact interval with $a \leqq x_0$, then any solution of the equation (8.19) existing at x_0 extends to $\langle x_0, b\rangle$.

PROOF. Let $|p(x)| + 1 \leqq M$ and $|q(x)| \leqq M$ on $\langle x_0, b\rangle$. Write the equation (8.19) in the vector form

$$\bar{y}' = f(x, \bar{y}), \quad \bar{y} = (y_1, y_2, y_3), \tag{8.20}$$

where $f_1(x, \bar{y}) = y_2$, $f_2(x, \bar{y}) = y_3$, $f_3(x, \bar{y}) = -q(x)y_1^r - p(x)y_2$. Then, to any solution y of (8.19) there corresponds a solution of the system (8.20), namely $\bar{y} = (y_1, y_2, y_3)$ with $y = y_1$, $y' = y_2$, $y'' = y_3$. Define $U(x, u) = M(u+1)$. Then $\|f(x, \bar{y})\| \leqq U(x, \|\bar{y}\|)$. The assertion of Theorem 8.4 now follows from a theorem by Wintner (Hartman [62], p. 29). □

THEOREM 8.5. Assume that $p(x) \geqq 0$, $q(x) \geqq 0$ and that there exists a continuous derivative $p'(x) \leqq 0$. Then every non-extendable solution of (8.19) has infinitely many zeros in a finite interval.

PROOF. Suppose that a non-extendable solution $y(x)$ of (8.19) exists and has only finitely many zeros in $\langle x_0, b)$, where $b<\infty$. Then, for some $x_1>x_0$, we have $y(x) \neq 0$ in $\langle x_1, b)$. Let $y(x)>0$ on $\langle x_1, b)$. Then $y'''(x)+p(x)y'(x) \leqq 0$ in $\langle x_1, b)$. Integrating the last inequality twice from x_1 to x, $x_1<x$, we see that $y(x)$ is bounded on $\langle x_1, b)$. Now, integrating the equation (8.19) twice between the limits x_1 and x, $x_1>x$, we find that also that $y'(x)$ and $y''(x)$ are bounded on $\langle x_1, b)$. Therefore $\lim_{x \to b^-} [(y(x))^2+(y'(x))^2+(y''(x))^2]<\infty$, and hence $y(x)$ can be extended beyond the point b (cf. Coddington and Lewinson [21], p. 61). □

We are now going to investigate the case $p(x) \leqq 0$ and $q(x) \leqq 0$.

LEMMA 8.2. Let $p(x) \leqq 0$ on $\langle a, \infty)$. Suppose that, on the same interval, $q(x)<0$ if $0<r<1$, and $q(x) \leqq 0$ if $r \geqq 1$. If $y(x)$ is a non-oscillatory solution of the equation (8.19), then there is a number $c \geqq a$ such that either $y(x)y'(x)>0$ for $x \geqq c$ or $y(x)y'(x) \leqq 0$ for $x \geqq c$.

PROOF. If $r \geqq 1$, Lemma 8.2 generalizes the results by Lazer [86] included in §3 of this monograph. The same argument as that used by A. C. Lazer proves Lemma 8.2 in the case $r \geqq 1$.

If $0<r<1$, assume that $y(x)>0$ for $x \geqq x_0$, $a \leqq x_0$. First of all, we prove that the zeros of the derivative $y'(x)$ are isolated in the interval $\langle x_0, \infty)$. Suppose the contrary, i.e. that some $x_1 \geqq x_0$ is a limit point of zeros of $y'(x)$. Then the continuity of $y'(x)$ implies that $y'(x_1)=0$. By Rolle's theorem, x is a limit point of zeros of $y''(x)$ as well. Thus $y''(x_1)=0$, by the continuity of $y''(x)$. Similarly, $y'''(x_1)=0$. On the other hand, since $q(x_1) \neq 0$ and $y(x_1)>0$, we get a contradiction of our assumption that $y(x)$ satisfies (8.19).

If $y'(x)$ has not more than one zero in (x_0, ∞), the assertion of the lemma is clear. If $y'(x)$ has two or more zeros in (x_0, ∞), we proceed as follows. Let $x_2<x_3$ be two neighbouring zeros of the derivative $y'(x)$ in

(x_0, ∞). Multiply the equation (8.19) by $y'(x)$ and integrate term by term from x_2 to x_3, obtaining

$$-\int_{x_2}^{x_3} (y''(t))^2\, dt + \int_{x_2}^{x_3} p(t)\,(y'(t))^2\, dt +$$

$$+\int_{x_2}^{x_3} q(t)\,(y(t))^r y'(t)\, dt = 0\,.$$

Since the first two terms in the last equality are non-positive, $q(x)<0$ and $y(x)>0$, we must have $y'(x)<0$ for $x\in(x_2, x_3)$. The zeros of $y'(x)$ being isolated, this argument shows that $y'(x)\leqq 0$ for $x\geqq x_2$. This proves Lemma 8.2. □

LEMMA 8.3. Let $f(x)$ be a function defined on $\langle x_0, \infty)$, $t_0\geqq 0$. Assume that $f(x)>0$ and that $f'(x)$ and $f''(x)$ exist for $x\geqq x_0$. Also, assume that $\lim_{x\to\infty} f(x)=A<\infty$ whenever $f'(x)\geqq 0$. Then

$$\liminf_{x\to\infty} |x^\alpha f''(x) - \alpha x^{\alpha-1} f'(x)| = 0$$

for any $\alpha\leqq 2$.

The proof is omitted (see Heidel [64], Lemma 2.2).

THEOREM 8.6. Let the coefficients of the differential equation (8.19) satisfy the hypotheses of Lemma 8.2. Moreover, let $-\infty < -M\leqq p(x)x^\alpha$ in $\langle a, \infty)$ and $\int_a^\infty t^\alpha q(t)\, dt = -\infty$ for some $\alpha\leqq 2$. If $y(x)$ is a non-oscillatory solution of (8.19) such that $y(x)y'(x)\leqq 0$, then $\lim_{x\to\infty} y(x)=0$.

PROOF. Let $y(x)>0$, which implies that $y(x)\leqq 0$ for $x\geqq x_0$. Assume that $\lim_{x\to\infty} y(x)=A>0$. Multiply (8.19) by x^α and integrate from x_0 and x, $x>x_0$. We obtain

$$[t^\alpha y''(t)]_{x_0}^x - [t^{\alpha-1}y'(t)]_{x_0}^x + \alpha(\alpha-1)\int_{x_0}^x t^{\alpha-2} y'(t)\, dt -$$

$$-M\int_{x_0}^{x} y'(t)\,dt+\int_{x_0}^{x} t^{\alpha}q(t)\,(y(t))^{r}\,dt \geqq 0\,. \tag{8.21}$$

Recall that both $\alpha(\alpha-1)\int_{x_0}^{x} t^{\alpha-2}y'(t)\,dt$ and $-M\int_{x_0}^{x} y'(t)\,dt$ are bounded for $x\to\infty$ since $y(x)$ has a finite limit and $\alpha\leqq 2$. Therefore (8.21) may be written in the form

$$x^{\alpha}y''(x)-\alpha x^{\alpha-1}y'(x)\geqq K-\int_{x_0}^{x} t^{\alpha}q(t)\,(y(t))^{r}\,dt\,, \tag{8.22}$$

where K is a finite constant. Since $\lim_{x\to\infty} y(x)=A>0$, the right-hand side of (8.22) diverges to infinity as $x\to\infty$. On the other hand, by Lemma 8.3 we have

$$\liminf_{x\to\infty} |x^{\alpha}y''(x)-\alpha x^{\alpha-1}y'(x)|=0\,.$$

This contradiction proves the theorem. □

The following two theorems may be established by similar, but somewhat more tedious proofs.

THEOREM 8.7. Let the hypotheses of Lemma 8.2 be met and let $\int_{a}^{\infty} tp(t)\,dt>-\infty$. If $y(x)$ is a non-oscillatory solution of (8.19) in $\langle x_0,\infty)$, then $y'(x)y(x)>0$ for $x\in\langle x_0,\infty)$.

THEOREM 8.8. Let the hypotheses of Lemma 8.2 be satisfied and let $-2/x^2\geqq p(x)\leqq 0$. If $y(x)$ is a non-oscillatory solution of (8.19), then $y(x)y'(x)>0$ in the respective interval.

For the proofs of the last two theorems, the reader is referred to Heidel [64].

For the sake of completeness, we give here a few more results (Heidel [64]) concerning the equation (8.19), closely related to those on the linear third order differential equation. The proofs will be omitted. The coefficients $p(x)$, $q(x)$ in the equation (8.19) will be assumed non-negative. In proving some of the results, an important role is played by the following lemma (Kiguradze [75]).

LEMMA 8.4. Let $f(x)$ be a continuous non-negative function defined on an interval $\langle x_0, \infty)$, $x_0 \geqq 0$. If $f^{(n)}(x) \leqq 0$, $n \geqq 2$, and $f^{(n-k)}(x) \geqq 0$, $k = 1, \ldots, n-1$, on $\langle x_0, \infty)$, then there are constants $A_k > 0$, $k = 1, \ldots, n-1$, such that for sufficiently large x we have

$$\frac{f(x)}{f^{(n-k)}(x)} \geqq A_k t \,.$$

The lemma is stated in a special form and is important for a large number of applications.

THEOREM 8.9. Let $p(x) \geqq 0$ and $q(x) \geqq 0$ for $a \leqq x < \infty$. Moreover, assume that

(i) $\displaystyle\int_a^\infty t^{2r} q(t)\, dt = \infty \quad$ if $\quad 0 < r < 1$;

(ii) $\displaystyle\int_a^\infty u(t) q(t)\, dt = \infty \quad$ if $\quad r = 1$,

where $u(x)$ is one of the functions $x^{2-\alpha}$, $x^2(\ln x)^{-1-\alpha}$, $x^2(\ln x)^{-1} \ln(\ln x)^{-1-\alpha}, \ldots$ for some $\alpha > 0$;

(iii) $\displaystyle\int_a^\infty t^{1+r} q(t)\, dt = \infty \quad$ if $\quad 1 < r$.

If $y(x)$ is a non-oscillatory solution of (8.19), then $|y(x)|$ is not non-decreasing.

The following lemma is a simple generalization of a lemma proved in (Lazer [86]) for the linear case.

LEMMA 8.5. Let $p(x) \geqq 0$, $q(x) \geqq 0$ and $p'(x) \leqq 0$ for $x \in \langle a, \infty)$. If $y(x)$ is a non-oscillatory (positive) solution of (8.19) and such that $F[y(c)] \geqq 0$ for some $c \in \langle a, \infty)$, that is,

$$F[y(t)] = (y'(x))^2 - 2y(x)y''(x) - p(x)\,(y(x))^2,$$

then there is $d \geqq c$ with

$$y(x) > 0, \quad y'(x) > 0, \quad y''(x) > 0 \quad \text{and} \quad y'''(x) \leqq 0 \quad \text{for } x \geqq d.$$

REMARK 8.1. In the linear case, the condition $p'(x) \leqq 0$ may be replaced by $2q(x) - p'(x) \geqq 0$, which is a condition on the Laguerre invariant.

Lemma 8.5 and Theorem 8.9 imply an interesting corollary analogous to an assertion about the linear equation.

COROLLARY 8.3. Let the hypotheses of Theorem 8.9 and those of Lemma 8.5 be met. Let $y(x)$ be an extendable solution of the equation (8.19), defined at $x_0 \geqq a$. Then $y(x)$ is non-oscillatory if and only if $F[y(x)] < 0$ for all $x \in \langle x_0, \infty)$.

PROOF. If $F[y(x)] < 0$ for $x \geqq x_0$, it is evident that $y(x)$ cannot have zeros at $x \geqq x_0$. Hence it is non-oscillatory.

Now assume that $F[y(x_1)] \geqq 0$ for some $x_1 \geqq x_0$. By Theorem 8.9 and Lemma 8.5, $y(x)$ is non-oscillatory and eventually positive. Suppose that $y(x)$ is non-oscillatory and eventually negative. Then $-y(x)$ is non-oscillatory and eventually positive. However, $F[-y(x_1)] = F[y(x_1)] \geqq 0$. This contradicts Theorem 8.9 and Lemma 8.5. Therefore, $y(x)$ is oscillatory. □

We shall state one more result, based on a well-known lemma of Nehari (Nehari [104]).

LEMMA 8.6. If $u'' + p(x)u = 0$ has a non-oscillatory solution in $\langle a, \infty)$, $p(x)$ is continuous in $\langle a, \infty)$ and $v(x)$ is any function in C^1 on $\langle b, c\rangle$, $a \leqq b < c < \infty$, such that $v(b) = 0$ and $v(t) \not\equiv 0$ on (b, c), then

$$\int_b^c (v'(t))^2 \, dt > \int_b^c p(t)\, (v(t))^2 \, dt .$$

THEOREM 8.10. Let $p(x) \geqq 0$, $q(x) > 0$ on $\langle a, \infty)$ and let the differential equation $u'' + p(x)u = 0$ have a non-oscillatory solution. If $y(x)$ is a non-oscillatory solution of (8.19), then there exists $d > a$ such that either $y(x)y'(x) \geqq 0$ for $x \geqq d$ or $y(x)y'(x) < 0$ for $x \geqq d$.

REMARK 8.2. A series of further results could be stated on the properties of solutions of the equation (8.19) and of other types of non-linear third order equations, having immediate applications to physical and engineering practice (Stewartson [134], Coppel [22], Heidel [65]).

§9. PHYSICAL AND ENGINEERING APPLICATIONS OF THIRD ORDER DIFFERENTIAL EQUATIONS

A number of physical and technological problems lead to the question of formulating a mathematical model describing a given process or a given structure (Marchenko [95]). In this connection, questions arise on the stability of the given process, on vibrations of the structure or on its critical load. From the mathematical point of view, these problems often lead to differential equations whose coefficients depend on a parameter, with certain additional (boundary) conditions, and the question arises how to determine the given parameter in order that the solution of the differential equation may describe the given process and also satisfy the respective additional conditions (the eigenvalue problem). In this part of the monograph, we give several examples in which a third order differential equation is applied to a physical or engineering problem.

However, such questions are often posed the other way round, that is to say, we are given parameter values and have to find the coefficients of a differential equation governing a given process. Mathematically speaking, this is the inverse problem for a linear differential operator, say, of order n,

$$L = D^n + q_{n-2}D^{n-2} + \ldots + q_0, \quad D = \frac{\mathrm{d}}{\mathrm{d}x}.$$

The task is to determine the coefficients of the operator L, knowing something about the spectrum of boundary-value problems or, on an infinite interval, utilizing our knowledge of the asymptotic properties of the eigenfunctions. Essentially, the problem is to determine which data are sufficient, and also to find a method for constructing the coefficients. This question is systematically treated in mathematical literature, especially in the case of the second order operator by Beals and Coifman [9, 10], Leibenzon [87], Matsaev [92] and others.

The topic is very important for applications, but due to its extent we shall not develop it in this monograph, although a number of new questions arise here about the third order differential equation, especially when a three-point boundary-value problem is considered.

1. *On Deflection of a Curved Beam*

We shall study the deflection of a curved beam having a constant or varying cross-section. The problem has been dealt with by many authors in the past. Our starting point will be a paper by Lockschin [91], whose mathematical formulation of the problem was based on the equilibrium equations (of the mathematical theory of elasticity)

$$\frac{dN}{ds}+Tk+X=0\,,\quad \frac{dT}{ds}-NK+Z=0\,,\quad \frac{dM}{ds}+N=0\,, \tag{9.1}$$

where N is the radial force, T the axial force, M the bending moment, k the curvature of the axis, X and Z are physical parameters. The bending moment M is related to the curvature k by

$$M=B(k-k_0)\,,$$

where B denotes the flexured rigidity (if the cross-section is constant, then B is constant) and k_0 is the initial curvature.

Introducing polar coordinates ϱ_0, Θ into (9.1) and assuming that

$$\frac{d}{d\Theta}(\varrho_0 X)-\varrho_0 Z=0\,,$$

A. Lockschin has derived the following differential equation for finding the critical values of the load X,

$$\left(\frac{d^2}{d\Theta^2}+1\right)\frac{1}{\varrho_0}\frac{dM}{d\Theta}+\frac{d}{d\Theta}\left(\frac{\varrho_0^2 XM}{B}\right)=0\,. \tag{9.2}$$

Suppose that the curved beam has the form of a circular arc with diameter $a(0\leqq\Theta\leqq\gamma,\ \gamma>0)$ and let it have a symmetry axis. Introduce a new independent variable, $\zeta=\Theta-\alpha$, $\alpha=\gamma/2$. In the case of an equally distributed load on the beam, $X=p$ (p being the force acting on a unit length), and the corresponding equation for M takes the form

$$\frac{1}{a^3}\left(\frac{d^2}{d\zeta^2}+1\right)\frac{dM}{d\zeta}+p\frac{d}{d\zeta}\left(\frac{M}{B}\right)=0\,. \tag{9.3}$$

If the beam has pin joints at the ends, then the boundary conditions (as derived by Lockschin [91]) have the form

$$M(-\alpha)=M(\alpha)=0, \int_{-\alpha}^{\alpha} \frac{M(\cos\zeta-\cos\alpha)}{B}\,d\zeta=0\,. \quad (9.4)$$

When solving the problem (9.3), (9.4), A. Lockschin proceeded as follows.

Any odd function which satisfies the equation (9.3) and the third condition of (9.4) necessarily satisfies the conditions for

$$\zeta=0\,, \quad M=0\,, \quad \frac{d^2M}{d\zeta^2}=0\,. \quad (9.5)$$

After the first integration of (9.3) we get

$$\frac{1}{a^2}\left(\frac{d^2}{d\zeta^2}+1\right)M+\frac{pM}{B}=c\,.$$

The conditions (9.5) imply that $c=0$. Thus we obtain the equation

$$\frac{B}{a^3}\left(\frac{d^2}{d\zeta^2}+1\right)M+pM=0 \quad (9.6)$$

with the boundary conditions

$$M(0)=M(\alpha)=0\,.$$

If the ends of the rod are fixed, then the boundary conditions on M are

$$\int_{-\alpha}^{\alpha}\frac{M}{B}\,d\zeta=0\,, \quad \int_{-\alpha}^{\alpha}\frac{M}{B}\cos\zeta\,d\zeta=0\,,$$

$$\int_{-\alpha}^{\alpha}\frac{M}{B}\sin\zeta\,d\zeta=0\,. \quad (9.7)$$

Taking (9.5) into account, we see that the first two of the conditions (9.7) will be fulfilled and we have to integrate the equation (9.6) under the constraint

$$M(0)=0\,, \quad \int_0^{\alpha}\frac{M}{B}\sin\zeta\,d\zeta=0\,.$$

For a beam having a constant cross-section, we get the boundary-value problem

$$\left(\frac{d^2}{d\zeta^2}+1\right)\frac{dM}{d\zeta}+\frac{pa^3}{B}\frac{dM}{d\zeta}=0\,, \tag{9.8}$$

$$M(\alpha)=M(-\alpha)=0\,,\quad \int_{-\alpha}^{\alpha}(\cos\zeta-\cos\alpha)M\,d\zeta=0\,. \tag{9.9}$$

The equation (9.8) has constant coefficients.

All the three boundary conditions will be met if we put

$$M=A\sin\frac{\pi\zeta}{\alpha}\,.$$

Substitute this in the differential equation (9.8). A condition is thus obtained for calculating the critical value of p,

$$p_{cr}=\frac{B}{a^3}\left(\frac{\pi^2}{\alpha^2}-1\right).$$

If the ends of the rod are fixed, the conditions (9.7) yield

$$\int_{-\alpha}^{\alpha}M\,d\zeta=0\,,\quad \int_{-\alpha}^{\alpha}M\cos\zeta\,d\zeta=0\,,$$

$$\int_{-\alpha}^{\alpha}M\sin\zeta\,d\zeta=0\,. \tag{9.10}$$

Now we suggest a solution in the form

$$M=A\sin m_1\zeta\,.$$

From (9.8) we get

$$p_{cr}=\frac{B(m_1^2-1)}{a^3}\,.$$

The first two of the conditions (9.10) are satisfied for every m_1. The third condition implies the following equation for m_1,

$$m_1\alpha\operatorname{cotg}m_1\alpha=\alpha\operatorname{cotg}\alpha\,.$$

We see that as far as the practical utilization of a result is concerned, it was always a method yielding concrete, explicit, and numerically computable results that proved effective.

Today in the computer era, it is sufficient to prove the existence and uniqueness of the solution of a given problem, to master a sufficiently exact numerical method, and to have at one's disposal a computer with a suitable software.

Therefore, let us return to the problem (9.3), (9.4). It can be written as follows:

$$y''' + \{[1+\lambda g(x)]y\}' = 0\,, \tag{9.11}$$

$$y(-l) = y(l) = 0\,, \quad \int_{-l}^{l} y(t)g(t)\,(\cos t - \cos l)\,\mathrm{d}t = 0\,. \tag{9.12}$$

Let $g(x) > 0$ be a continuous function of $x \in \langle -l, l\rangle$.
Integrate the equation (9.11) from $-l$ to $x \in (-l, l\rangle$. We get

$$y'' + y + \lambda g(x)y = y''(-l)\,.$$

Put $y''(-l) = 0$ and multiply the last equation by $\cos x - \cos l$. We get

$$-(y''+y)\,(\cos x - \cos l) = \lambda y g(x)\,(\cos x - \cos l)\,,$$

$$-\int_{-l}^{l} (y''(t) + y(t))\,(\cos t - \cos l)\,\mathrm{d}t =$$

$$= \lambda \int_{-l}^{l} y(t)g(t)\,(\cos t - \cos l)\,\mathrm{d}t\,.$$

Taking into account the first two of the conditions (9.12), we find that the integral on the left-hand side of the last equation becomes:

$$-\int_{-l}^{l} [y''(t) + y(t)]\,(\cos t - \cos l)\,\mathrm{d}t = \cos l \int_{-l}^{l} y(t)\,\mathrm{d}t\,.$$

Thus the third condition (9.12) will be satisfied if $l = \pi/2 + k\pi$, $k = 0, 1, 2, \dots$ and, of course, if $y''(-l) = 0$.

We have just proved the following lemma.

LEMMA 9.1. Any solution y of the differential equation (9.11) satisfying the first two of the conditions (9.12) satisfies also the third condition whenever

$$y''(-l) = 0, \quad l = \frac{\pi}{2} + k\pi\,, \quad k = 0, 1, 2, \dots\,.$$

COROLLARY 9.1. For the deflection of a beam having the form of a circular arc Lemma 9.1 implies that the arc must be the semicircumference $l=\alpha=\pi/2$.

COROLLARY 9.2. The boundary problem to be solved is

$$y'''+\{[1+\lambda g(x)]y\}'=0\,,$$

$$y\left(-\frac{\pi}{2}\right)=y''\left(-\frac{\pi}{2}\right)=0\,,\quad y\left(\frac{\pi}{2}\right)=0\,. \tag{9.13}$$

Integrating (9.11) from $-\pi/2$ to x we get

$$y''+[1+\lambda g(x)]y=0 \tag{9.14}$$

$$y\left(-\frac{\pi}{2}\right)=y\left(\frac{\pi}{2}\right)=0\,. \tag{9.15}$$

Now (9.14), (9.15) is a Sturm-Liouville boundary-value problem. Its first eigenvalue is $\lambda_0=0$ and the corresponding eigenfunction is $y_0(x)=\cos x$, which, however, is not significant as regards the beam. The other eigenvalues are positive and form an increasing sequence $\{\lambda_\nu\}_{\nu=1}^{\infty}$. With respect to the beam, the significant eigenvalue is λ_1, and perhaps some further eigenvalues which may represent a critical load of the beam.

REMARK 9.1. The third of the boundary conditions (9.12) is in an integral form and to meet it we have assumed that $l=\pi/2$, that is, the beam has the form of a semicircumference. This, however, is a restriction. Therefore, the problem for $0<l<\pi/2$ remains open.

From the mathematical point of view, the problem (9.11), (9.12) may be generalized as follows.

Consider the differential equation

$$y'''+[1+\lambda g(x)]y'+\lambda h(x)y=0 \tag{9.16}$$

and the boundary conditions

$$y(-l)=y(l)=0\,,\quad \int_{-l}^{l}(\cos t-\cos l)\left[g(t)y(t)+\right.$$

$$\left.+\int_{-1}^{t}[h(\tau)-g'(\tau)]\,y(\tau)\,d\tau\right]dt=0\,, \tag{9.17}$$

with $g(x)>0$, $g'(x)\geqq 0$, $h(x)\geqq 0$ being continuous functions of $x\in\langle -l, \infty)$, and let $h(x)-1/2g'(x)\geqq 0$ for $x\in\langle -l, \infty)$, where the equality holds at isolated points only. $\lambda\geqq 0$ is a parameter.

If $h(x)=g'(x)$, the problem (9.16), (9.17) reduces to (9.11), (9.12) as a particular case (except the assumption of continuity of $g'(x)\geqq 0$).

REMARK 9.2. As in the case of the problem (9.11), (9.12) it can be proved similarly that the third of the conditions (9.17) is satisfied whenever $l=\pi/2+k\pi$, $k=0, 1, 2, \ldots$, and $y''(-l)=0$, because the equation (9.16) may be written in the form

$$y'''+\{[1+\lambda g(x)]y\}'+\lambda[h(x)-g'(x)]y=0\,.$$

Thus, instead of (9.17), we have to consider the conditions

$$y(-l)=y''(-l)=0,\quad y(l)=0,\quad l=\pi/2+k\pi,$$
$$k=0, 1, 2, \ldots\,.$$

In the sequel, we shall investigate the case $k=0$, that is, $l=\pi/2$.

REMARK 9.3. It follows from Lemma 4.2 that if the coefficients of the differential equation (9.16) satisfy the above assumptions and $y(x, \lambda)$ is a non-trivial solution of (9.16) with $y(-l, \lambda)=0$, then the zeros of $y(x, \lambda)$ (if they exist) in $(-l, \infty)$ are continuous functions of the parameter $\lambda\in\langle 0, \infty)$.

Using Remark 9.3, Theorem 2.16 and oscillation Theorem 4.5 b) we shall prove the following theorem.

THEOREM 9.1. Let the coefficients of the differential equation (9.16) satisfy the above assumptions. Then there exist a sequence $\{\lambda_\nu\}_{\nu=1}^{\infty}$ of values of the parameter λ and a sequence of functions $\{y_\nu\}_{\nu=1}^{\infty}$ such that $y_\nu=y(x, \lambda_\nu)$ is a solution of (9.16) satisfying the boundary conditions

$$y\left(-\frac{\pi}{2}, \lambda\right)=y''\left(-\frac{\pi}{2}, \lambda\right)=0,\quad y\left(\frac{\pi}{2}, \lambda\right)=0 \qquad (9.18)$$

and $y(x, \lambda_\nu)$ has exactly ν zeros in $(-\pi/2, \pi/2)$.

PROOF. First of all, we should observe that the coefficients of (9.16) satisfy the hypotheses of Lemma 4.2, Theorem 2.16 and oscillation Theorem 4.5 b) for $x \in \langle -\pi/2, \infty)$ and $\lambda \in \langle 0, \infty)$. For $\lambda = 0$, there is a solution of (9.16), $y_0 = y(x, 0) = \cos x$, satisfying (9.18) and having no other zero in $(-\pi/2, \pi/2)$.

Denote by $x_\nu(\lambda)$, $\nu = 1, 2, \ldots$, the zeros of $y(x, \lambda)$ to the right of $-\pi/2$ satisfying $y(-\pi/2, \lambda) = y''(-\pi/2, \lambda) = 0$ for every $\lambda \in \langle 0, \infty)$. The above reasoning yields $x_1(0) = \pi/2$ and $x_2(0) > \pi/2$. By Theorem 4.5 b), there is $\bar{\lambda} > 0$ with $x_2(\bar{\lambda}) < \pi/2$. According to Remark 9.3, $x_2(\lambda)$ is a continuous function of λ and hence there exists $\lambda_1 \in (0, \bar{\lambda})$ with $x_2(\lambda_1) = \pi/2$. Thus the solution $y_1 = y(x, \lambda_1)$ of (9.16) satisfies the boundary conditions (9.18) and has exactly one zero in $(-\pi/2, \pi/2)$.

Obviously, $x_3(\lambda_1) > \pi/2$. In virtue of Theorem 4.3 b), there exists $\bar{\bar{\lambda}} > \lambda_1$ such that $x_3(\bar{\bar{\lambda}}) < \pi/2$. The continuity of $x_3(\lambda)$ with respect to λ implies the existence of a $\lambda_2 \in (\lambda_1, \bar{\bar{\lambda}})$ such that $x_3(\lambda_2) = \pi/2$ and the solution $y_2 = y(x, \lambda_2)$ of (9.16) satisfies (9.18) and has exactly two zeros in $(-\pi/2, \pi/2$. Proceeding in this way, we prove the existence of $\pi/2$ sequences $\{\lambda_\nu\}_{\nu=1}^{\infty}$ and $\{y_\nu\}_{\nu=1}^{\infty}$. Thus the theorem is proved. □

REMARK 9.4. In the case of a beam having a parabolic shape and a constant cross-section with an equally distributed load of intensity p acting parallel to the symmetry axis of the beam we have

$$B = \text{const}, \quad X = p \cos^2 \zeta, \quad \varrho_0 = \frac{a}{\cos^2 \zeta},$$

where a is the radius of curvature at $\zeta = 0$. Putting these expressions in the equation (9.2), we get

$$\left(\frac{d^2}{d\zeta^2} + 1\right) \cos^3 \zeta \frac{dM}{d\zeta} + \frac{pa^3}{B} \frac{d}{d\zeta} \left(\frac{M}{\cos^4 \zeta}\right) = 0 . \tag{9.19}$$

The constraints for the solution of this equation are

$$M(-\alpha) = M(\alpha) = 0 , \quad \int_{-\alpha}^{\alpha} \left[\frac{1}{\cos^2 \alpha} - \frac{1}{\cos^2 \zeta}\right] \frac{M}{\cos^3 \zeta} d\zeta = 0 . \tag{9.20}$$

The method used by Lockschin [91] for solving the problem (9.19), (9.20) will not be presented here.

REMARK 9.5. In the case where the beam has not a constant cross-section and

$$B=\frac{B_0}{\cos^3\zeta},$$

the problem (9.19), (9.20) reduces to

$$\left(\frac{d^2}{d\zeta^2}+1\right)\cos^3\zeta\,\frac{dM}{d\zeta}+\frac{pa^3}{B_0}\frac{d}{d\zeta}\left(\frac{M}{\cos\zeta}\right)=0,$$

$$M(-\alpha)=M(\alpha)=0,\quad \int_{-\alpha}^{\alpha}\frac{M}{\cos^2\alpha-\cos^2\zeta}\,d\zeta=0.$$

REMARK 9.6. From the mathematical point of view, it is interesting that in the above cases one of the boundary conditions is expressed by an integral and the theory of the third order differential equation can be applied.

2. *Three-layer Beam*

A three-layer beam is formed by parallel layers of different materials. For an equally loaded beam of this type, Krajcinovic [81] has shown that the deflection ψ is governed by an ordinary third order linear differential equation

$$\frac{d^3\psi}{dx^3}-k^2\frac{d\psi}{dx}+a=0, \tag{9.21}$$

where k^2 and a are physical parameters depending on the elasticity of the layers. The condition of zero moment at the free ends implies the boundary conditions

$$\frac{d\psi(0)}{dx}=\frac{d\psi(1)}{dx}=0 \tag{9.22}$$

and symmetry yields the third boundary condition

$$\psi\left(\frac{1}{2}\right)=0\,. \tag{9.23}$$

Although the formulation of this problem is a simple application of the minimal potential energy principle, the details of deduction are more complicated and therefore will not be given here.

We shall present one of the methods (Na [103]) (the superposition method) for guaranteeing the existence of a solution to the boundary-value problem (9.21), (9.22), (9.23). This method consists in showing that the problem (9.21), (9.22), (9.23) is equivalent to three initial value problems at 0, 1, 1/2.

Let us postulate a solution of the problem in the form

$$y(x)=y_1(x)+\mu y_2(x)+\lambda y_3(x)\,,$$

where λ, μ are constants to be determined later and y_1, y_2, y_3 are suitable functions.

Substituting for y in (9.21), we get

$$y_1''''-k^2y_1'+a+\mu(y_2''''-k^2y_2')+\lambda(y_3''''-k^2y_3')=0\,.$$

In order that y should satisfy (9.21) we choose y_1, y_2, y_3 so that

$$\begin{aligned}&y_1''''-k^2y_1'+a=0\,,\\&y_1(0)=0,\quad y_1'(0)=0\,,\quad y_1''(0)=0\,,\end{aligned} \tag{9.24}$$

also

$$\begin{aligned}&y_2''''-k^2y_2'=0\,,\\&y_2(0)=1\,,\quad y_2'(0)=0\,,\quad y_2''(0)=0\,,\end{aligned} \tag{9.25}$$

and

$$\begin{aligned}&y_3''''-k^2y_3'=0\,,\\&y_3(0)=0\,,\quad y_3'(0)=0\,,\quad y_3''(0)=1\,.\end{aligned} \tag{9.26}$$

Then y will evidently be a solution of (9.21) satisfying the first of the conditions (9.22). It remains to determine μ and λ so that the second and third conditions of (9.22) may be satisfied as well. The following conditions for μ, λ result:

$$y_1\left(\frac{1}{2}\right)+\mu y_2\left(\frac{1}{2}\right)+\lambda y_3\left(\frac{1}{2}\right)=0\,,$$

$$y_1'(1) + \mu y_2'(1) + \lambda y_3'(1) = 0 .$$

Clearly, such μ and λ can be found provided that the determinant $y_2(1/2)y_3'(1) - y_2'(1)y_3(1/2) \neq 0$, and that at least one of the numbers $y_1(1/2)$, $y_1'(1)$ is non-zero.

It can be calculated explicitly that

$$\psi \equiv y = \frac{a}{k^3}\left[\sin\frac{k}{2} - shkx\right] + \frac{a}{k^2}\left(x - \frac{1}{2}\right) +$$

$$+ \frac{a}{k^3}\, th\,\frac{k}{2}\left[chkx - ch\,\frac{k}{2}\right].$$

3. *Survey of Some Other Applications of Third Order Differential Equations*

a) An electromagnetic wave incident on a system of charges sets them into motion. This motion is accompanied by radiation in all directions. The wave is dispersed and the dispersion can be appropriately characterized by the ratio of the flux density of the energy radiated by the dispersing system in a given direction to that of the flux density of the radiation energy incident on the system. This ratio has the dimension of a surface and is called the scattering cross-section. The topic is discussed in detail in the book by Landau and Lifshits [84].

In the case of one charge performing small oscillations with an angular frequency w_0, taking the brake force of radiation into account, the equations governing this oscillator take the form

$$\ddot{r} + w_0^2 r = \frac{e}{m} E_0 \exp(-iwt) + \frac{2e^2}{3mc^3}\dddot{r}, \tag{9.27}$$

where r is the vector of forced vibrations and e, m, E_0, c are physical quantities. The last term in the equation represents the brake force of radiation. Putting (approximately) $\dddot{r} = -w_0^2\dot{r}$, the equation (9.27) reduces to

$$\ddot{r} + \gamma\dot{r} + w_0^2 r = \frac{e}{m}\exp(-iwt), \quad \gamma = \frac{2e^2}{3mc^3} w_0^2 .$$

Hence we deduce for the forced vibrations that

$$r = \frac{e}{m} E_0 \frac{\exp(-iwt)}{w_0^2 - w^2 - iw\gamma}$$

and this result enables us to calculate the dispersion of the incident wave (Landau and Lifshits [84]).

However, the equation (9.27) is of the third order and the difficulty of solving it was by-passed by lowering its order.

b) Problems of regulation and control of some actions by a control lever or by a signal reduce to solving third order equations (Rocard [120]).

As an example, we discuss the regulation of a steam turbine.

The motion $x(t)$ of a steam supply control slide valve is governed by the third order differential equation

$$m \frac{d^3x}{dt^3} + f \frac{d^2x}{dt^2} + k \frac{dx}{dt} + h \frac{\alpha}{I} x = 0 ,$$

where m is the mass of the valve, f is friction, k is a constant characterizing the properties of the slide valve spring, h is a constant depending on the dimensions of the equipment, α is a proportionality constant relating the motion and the acceleration of the control valve, and I is the moment of inertia of the turbine.

REMARK 9.7. Another application of third order differential equation theory is encountered in the control of a flying apparatus in cosmic space. However, in this case the analysis and control of the flight requires deeper knowledge of the control theory (Petrov et al. [112]).

References

[1] Ahmad, S. and Lazer, A. C. (1969), On the Oscillatory Behaviour of a Class of Linear Third Order Differential Equations, *J. Math. Anal. Appl.*, **28**, 681—689.

[2] Appel, P. (1880), Sur la transformation des équations différentielles linéaires, *C. R. Séanc. Acad. Sci. (Paris)*, **91**, 211—214.

[3] Ascoli, G. (1940), Sulla decomposizione degli operatori differenziali lineari e sopra alcune questioni geometriche che vi sí connettono, *Rev. Math. Fis. Teor.*, Ser. A, 189—215.

[4] Azbelev, N. V. (1958), On the question of the Estimation of the Number of Zeros of Solutions of the Equation $y''' + p(x)y' + q(x)y = 0$, *Nauch. Dokl. Vyssh. Shkoly, Fiz.-Mat. Nauki*, **3**, 4—5.

[5] Azbelev, N. V. and Tsalyuk, E. B. (1960), On the Question of the Distribution of Zeros of Solutions of a Third Order Differential Equation, *Mat. Sbor.*, **51**, 475—485 (in Russian).
Азбелев, Н. В. и Цалюк, Э. Б. (1960), К вопросу о распределении нулей решений линейного дифференциального уравнения третьего порядка, *Мат. сбор.*, **51**, 475—485.

[6] Baar, D. and Sherman, T. (1973), Existence and Uniqueness of Solutions of Three-point Boundary Value Problems, *J. Diff. Eqs.*, **13**, 197—212.

[7] Barrett, J. H. (1964), Canonical Forms for Third Order Linear Differential Equation, *Ann. Mat. Pura Appl.*, **4**, 253—274.

[8] Barrett, J. H. (1968), Third Order Differential Equations with Nonnegative Coefficients, *J. Math. Anal. Appl.*, **24**, 212—224.

[9] Beals, R. and Coifman, R. (1980—1982): Scattering, transformations spectrales, et équations d'évolution non linéaires, *Séminaire Goulavuic—Meyer—Schwartz*, 1980—1981, exp. 22; 1981—1982, exp. 21, Ecole polytechnique, Palaiseau.

[10] Beals, R. and Coifman, R. (in press), Scattering and Inverse Scattering for First Order Systems, *Commun. Pure Appl. Math.*

[11] Bespalova, S. and Klokov, E. (1976), Three-point Boundary Value Problem for Ordinary Third Order Differential Equation, *Diff. Uravneniya*, **12**, 963—970 (in Russian).
Беспалова, С. и Клоков, Е. (1976), Трехточечная проблема граничного

значения для обыкновенного дифференциального уравнения третьего порядка, *Диф. уравнения*, **12**, 963—970.

[12] Birkhoff, G. D. (1910), On the Solutions of Ordinary Linear Homogeneous Differential Equations of the Third Order, *Ann. Math.*, **2**, 103—127.

[13] Borůvka, O. (1953), On Oscillating Solutions of Second Order Linear Differential Equations, *Czechoslov. Math. J.*, **3**, 199—255.

[14] Borůvka, O. (1956), Sur la transformation des intégrales des équations différentielles linéaires ordinaires du second ordre, *Ann. Mat. Pura Appl.*, **41**, 325—342.

[15] Borůvka, O. (1957), Théorie analytique et constructive des transformations différentielles linéaires du second ordre, *Bull. Math. Soc. Sci. Math. Phys. Rep. Pop. Roum.*, **49**, 125—130.

[16] Borůvka, O. (1960), Sur les transformations différentielles linéaires complètes du second ordre, *Ann. Mat. Pura Appl.*, **49**, 229—252.

[17] Borůvka, O. (1962), Sur la structure de l'ensemble des transformations différentielles linéaires complétes du second ordre, *Ann. Mat. Pura Appl.*, **58**, 317—334.

[18] Borůvka, O. (1967), *Lineare Differentialtransformationen 2. Ordnung*, VEB Deutscher Verlag der Wissenschaften, Berlin.

[19] Boulanger, J. (1939), Sur l'équation différentielle du troisième ordre, *Bull. Soc. Roy. Sci. (Liège)*, **8**, 208—223.

[20] Červeň, J. (1964), On a Sufficient Condition for Non-oscillatoricity of Solutions of the Third Order Linear Differential Equation (in Slovak) — O jednej postačujúcej podmienke neoscilatoričnosti riešení lineárnej diferenciálnej rovnice tretieho rádu, *Acta Fac. R. N. Univ. Comenianae, Math.*, **9**, 63—69.

[21] Coddington, E. A. and Lewinson, N. (1965), *Theory of Ordinary Differential Equations*, McGraw-Hill Inc., New York.

[22] Coppel, W. (1960), On a Differential Equation of Boundary Layer Theory, *Phil. Trans. Roy. Soc.*, **253**, 101—136.

[23] Dolan, J. M. (1970), On the Relationship between the Oscillatory Behavior of a Linear Third Order Differential Equation and Its Adjoint, *J. Diff. Eqs.*, **7**, 367—388.

[24] Gallina, G. (1933), Sull'integrazione delle equazioni differenziali lineari omogenee autoaggiunte, *Rc. Res. Ist. Lombardo Sci. Lett.*, **2**, 724—730.

[25] Gallina, G. (1933), Su una classe di equazioni differenziali lineari omogenee del terzo ordine, *Boll. U. Mat. Ital.*, **12**, 142—145.

[26] Gera, M. (1970), Bedingungen der Nicht-oszillationsfähigkeit für die lineare Differentialgleichung dritter Ordnung $y''' + p_1(x)y'' + p_2(x)y' + p_3(x)y = 0$, *Acta Fac. R. N. Univ. Comenianae, Math.*, **24**, 145—158.

[27] Gera, M. (1970), Allgemeine Bedingungen der Nicht-oszillationsfähigkeit für die lineare Differentialgleichung dritter Ordnung $y''' + p_1(x)y'' + p_2(x)y' + p_3(x)y = 0$, *Mat. Čas.*, **20**, 49—61.

[28] Gera, M. (1971), Einige oszillatorische Eigenschaften der Lösungen der Differentialgleichung dritter Ordnung $y''' + p(y)y' + q(x)y = 0$, *Scr. Fac. Sci. Mat. UJEP (Brno), Arch. Math.*, **2**, 65—76.

[29] Gera, M. (1971), Nichtoszillatorische und oszillatorische Differentialgleichungen dritter Ordnung, *Čas. Pěst. Mat.*, **96**, 278—293.

[30] Gera, M. (1971), Bedingungen der Nicht-oszillationsfähigkeit und der Oszillationsfähigkeit für die lineare Differentialgleichung dritter Ordnung, *Mat. Čas.*, **21**, 65—80.

[31] Gera, M. (1974), Über einige Eigenschaften der Lösungen der Gleichung $x''' + a(t)x'' + b(t)x' + c(t)x = 0$, $c(t) \geqq 0$, *Mat. Čas.*, **24**, 357—370.

[32] Gera, M. (1975), Einige Integralbedingungen für das Nichtoszillieren der linearen Differentialgleichungen dritter Ordnung, *Acta Fac. R. N. Univ. Comenianae, Math.*, **31**, 73—102.

[33] Gera, M. (1976), On the Behaviour of Solutions of the Differential Equation $x''' + a(t)x' + b(t)x' + c(t)x = 0$, *Habilitation Thesis* (in Slovak) — O chovaní sa riešení diferenciálnej rovnice $x''' + a(t)x'' + b(t)x' + c(t)x = 0$, Faculty of Mathematics and Physics, Comenius University, Bratislava.

[34] Gera, M. (1978), Integralbedingungen für das Nichtoszillieren der Lösungen der linearen Differentialgleichung dritter Ordnung, *Acta Fac. R. N. Univ. Comenianae, Math.*, **30**, 35—50.

[35] Gerová, M. (1974), A Sufficient Disconjugacy Condition for the Third Order Differential Equation, *Mat. Čas.*, **24**, 253—258.

[36] Ghizzetti, A. (1949), Un teorema sul comportamento asintotico degli integrali delle equazioni differenziali omogenee, *Rc. Mat. Appl.*, **5**, 28—42.

[37] Ghizzetti, A. (1964), *Lezioni sui procedimenti di quazilinearizzazione*, CIME, Perugia.

[38] Greguš, M. (1955), On Certain Properties of Solutions of the Third Order Linear Homogeneous Differential Equation (in Slovak) — O niektorých vlastnostiach riešení lineárnej homogénnej diferenciálnej rovnice tretieho rádu, *Mat.-Fyz. Čas.*, **2**, 73—85.

[39] Greguš, M. (1955), On Some New Properties of Solutions of the Differential Equation $y''' + Qy' + Q'y = 0$ (in Slovak) — O niektorých nových vlastnostiach riešení diferenciálnej rovnice $y''' + Qy' + Q'y = 0$, *Spisy Přír. Fak. MU (Brno)*, **365**, 1—18.

[40] Greguš, M. (1956), On Some Relations between the Integrals of Mutually Adjoint Third Order Linear Differential Equations and on a Boundary Value Problem (in Slovak) — O niektorých vzťahoch medzi integrálmi navzájom adjungovaných lineárnych diferenciálnych rovníc tretieho rádu a o jednom okrajovom probléme, *Acta Fac. R. N. Univ. Comenianae*, **1**, 262—272.

[41] Greguš, M. (1957), On Some New Boundary-value Problems of the Third Order Differential Equation, *Czechoslov. Math. J.*, **7**, 41—47 (in Russian).
Грегуш, М. (1957), О некоторых новых краевых задачах дифференциального уровнения третьего порядка, Чехослов. мат., **7**, 41—47.

[42] Greguš, M. (1957), On Third Order Linear Differential Equation with Constant Coefficients (in Slovak) — O lineárnej diferenciálnej rovnici tretieho rádu s konštantnými koeficientmi, *Acta Fac. R. N. Univ. Comenianae*, **2**, 61—65.

[43] Greguš, M. (1957), Homogeneous Boundary Value Problem for Integrals of the Third Order Linear Differential Equation (in Slovak) — Homogénny okrajový problém pre integrály lineárnej diferenciálnej rovnice tretieho rádu, *Acta Fac. R. N. Univ. Comenianae*, **5—6**, 219—228.

[44] Greguš, M. (1958), A Note on the Oscillation Properties of Solutions of the Linear Third Order Differential Equation (in Slovak) — Poznámka k oscilatorickým vlastnostiam riešení lineárnej diferenciálnej rovnice tretieho rádu, *Acta Fac. R. N. Univ. Comenianae, Math.*, **1**, 23—28.

[45] Greguš, M. (1959), Oszillatorische Eigenschaften der Lösungen der linearen Differentialgleichung dritter Ordnung $y''' + 2Ay' + (A' + b)y = 0$, wo $A = A(x) \leqq 0$ ist, *Czechoslov. Math. J.*, **9**, 416—428.

[46] Greguš, M. (1962), Über einige Eigenschaften der Buschel von Lösungen der linearen Differentialgleichungen dritter Ordnung, *Čas. Pěst. Mat.*, **87**, 311—319.

[47] Greguš, M. (1963), Über die lineare homogene Differentialgleichung dritter Ordnung, *Wiss. Z. Univ. Halle, Math.-Naturwiss.*, **12**, 265—286.

[48] Greguš, M. (1963), Über die asymptotischen Eigenschaften der Lösungen der linearen Differentialgleichung dritter Ordnung, *Ann. Mat. Pura Appl.*, **4**, 1—10.

[49] Greguš, M. (1963), Bemerkungen zu den unlösbaren Randwertproblemen dritter Ordnung, *Acta Fac. R. N. Univ. Comenianae, Math.*, **7**, 639—647.

[50] Greguš, M. (1964), Über das Randwertproblem der n-ten Ordnung in m Punkten, *Acta Fac. R. N. Univ. Comenianae, Math.*, **9**, 49—55.

[51] Greguš, M. (1965), Über die Eigenschaften der Lösungen einiger quasi-linearer Gleichungen dritter Ordnung, *Acta Fac. R. N. Univ. Comenianae, Math.*, **10**, 11—22.

[52] Greguš, M. (1966), On Oscillatoricity of Solutions of Third Order Linear Differential Equation (in Slovak) — O oscilatoričnosti riešení lineárnej diferenciálnej rovnice tretieho rádu, in Sborník družby pěti bratských univerzit, Kyjev, Krakov, Debrecín, Bratislava, Brno, 146—150.

[53] Greguš, M. (1972), Three-point Boundary Value Problem in a Differential Equation of the Third Order, *Proc. Equadiff 3* (Brno), 115—118.

[54] Greguš, M. (1974), Remarks on a Three-point Boundary Value Problem in a Differential Equation of the Third Order, *Ann. Pol. Math.*, **29**, 229—232.

[55] Greguš, M. (1975), On a Special Boundary Value Problem of G. Sansone, *Boll. U. Mat. Ital.*, **4**, 344—348.

[56] Greguš, M. (1978), On a Special Criterion on the Disconjugacy of a Linear Differential Equation of the Third Order, *Proc. Equadiff 78* (Firenze), 237—244.

[57] Greguš, M. and Abdel Karim, R. I. I. (1967), Some Properties of Some Special Differential Equations of the Third Order, *Proc. Math. and Phys. Soc., U.A.R.*, **31**, 67—74.

[58] Greguš, M. and Abdel Karim, R. I. I. (1968), Boundedness of the Solutions of the Differential Equation $(py')'' + (py')' + ry = 0$, *Proc. Math. and Phys. Soc., U.A.R.*, **32**, 107—110.

[59] Greguš, M. and Abdel Karim R. I. I. (1969), Bands of Solutions of Some Special

Differential Equations of the Third Order, *Acta Fac. R. N. Univ. Comenianae, Math.*, **22**, 57—66.

[60] Greguš, M., Neuman, F. and Arscott, F. M. (1971), Three-point Boundary Value Problems in Differential Equations, *J. London Math. Soc.*, **2**, 429—436.

[61] Hanan, M. (1961), Oscillation Criteria for Third Order Linear Differential Equations. *Pac. J. Math.*, **11**, 916—944.

[62] Hartman, P. (1964), *Ordinary Differential Equations*, Wiley and Sons, New York.

[63] Hartman, P. and Wintner, A. (1945), Asymptotic Integrations of Linear Differential Equations, *Amer. J. Math.*, **66**, 45—86.

[64] Heidel, J. W. (1968), Qualitative Behavior of Solutions of a Third Order Nonlinear Differential Equation, *Pac. J. Math.*, **27**, 507—526.

[65] Heidel, J. W. (1973), A Third Order Differential Equation Arising in Fluid Mechanics, *Z. Angew. Math. Mech.*, **53**, 167—170.

[66] Hustý, Z. (1965—1966), Über die Transformation und Äquivalenz linearer Differentialgleichungen von höherer als der zweiten Ordnung, *Czechoslov. Math. J.*, I. Teil — **15**, 479—502, II. Teil — **16**, 1—13, III. Teil — **16**, 161—185.

[67] Jackson, L. (1973), Existence and Uniqueness of Solutions of Boundary Value Problems for Third Order Differential Equations, *J. Diff. Eqs.*, **13**, 432—437.

[68] Jackson, L. and Schrader, K. (1971), Existence and Uniqueness of Solutions of Boundary Value Problems for Third Order Differential Equations, *J. Diff. Eqs.*, **9**, 46—54.

[69] Jones, G. D. (1970), A property of $y'''+p(x)y'+1/2p'(x)y=0$, *Proc. Amer. Math. Soc.*, **26**, 273—276.

[70] Jones, G. D. (1973), Oscillation Properties of Third Order Differential Equations, *Rocky Mt. J. Math.*, **3**, 507—513.

[71] Jones, G. D. (1973), An Asymptotic Property of Solutions of $y'''+p(y)y'+q(x)y=0$, *Pac. J. Math.*, **47**, 135—138.

[72] Jones, G. D. (1974), Oscillatory Behavior of Third Order Differential Equations, *Proc. Amer. Math. Soc.*, **43**, 133—136.

[73] Jones, G. D. (1974), Properties of Solutions of a Class of Third Order Differential Equations, *J. Math. Anal. Appl.*, **48**, 165—169.

[74] Jones, G. D. (1976), Oscillation Criteria for Third Order Differential Equations, *Siam.J. Math. Anal.*, **7**, 13—15.

[75] Kiguradze, I. T. (1962), Oscillation Criteria for Third Order Linear Differential Equations, *Dokl. AN SSSR*, **114**, 649—652 (in Russian).
Кигурадзе, И. Т. (1962), Критерии осцилляции для линейных дифференциальных уравнений третьего порядка, *Докл. АН СССР* **114**, 649—652.

[76] Kiguradze, I. T. (1965), Concerning the Question of Oscillatory Solutions of Non-linear Differential Equations, *Diff. Uravneniya*, **1**, 995—1006 (in Russian).
Кигурадзе, И. Т. (1965), К вопросу осцилляционных решений нелинейных дифференциальных уравнений, *Диф. уравнения*, **1**, 995—1006.

[77] Kiguradze, I. T. (1975), *Some Singular Boundary Value Problems for Ordinary Differential Equations*, Izd. Tbil. Univ., Tbilisi (in Russian).

Кугурадзе, И. Т. (1975), *Некоторые проблемы сингулярной граничной задачи для обыкновенных дифференциальных уравнений*, Изд. Тбил. унив., Тбилиси.

[78] Klassen, G. (1971), Differential Inequalities and Existence Theorems for Second and Third Order Boundary Value Problems, *J. Diff. Eqs.*, **10**, 529—537.

[79] Kondratev, V. A. (1958), Zeros of Solutions of the Equation $y^{(n)}+p(x)y=0$, *Dokl. AN SSSR*, **120**, 1180—1182 (in Russian).
Кондратьев, В. А. (1958), Нулевые точки решений уравнения $y^{(n)}+p(x)y=0$, *Докл. АН СССР*, **120**, 1180—1182.

[80] Kondratev, V. A. (1959), On Oscillatoricity of Solutions of Third and Fourth Order Linear Equations, *Trudy Mosk. Mat. Obshch.*, **8**, 259—281 (in Russian).
Кондратьев, В. А. (1959), О колеблемости решений линейных уравнений третьего и четвертого порядка, *Труды Моск. мат. общ.*, **8**, 259—281.

[81] Krajcinovic, D. (1972), Sandwich Beam Analysis, *J. Appl. Mech.*, **39**, 773—778.

[82] Laguerre, E. (1879), Sur les equations différentielles linéaires du troisième ordre, *C. R. Séanc. Acad. Sci. (Paris)*, **88**, 116—119.

[83] Laitoch, M. (1960), Transformation of the Solutions of Linear Differential Equations, *Czechoslov. Math. J.*, **10**, 258—270 (in Russian).
Лайтох, М. (1960), Трансформация решений линейных дифференциальных уравнений, *Czechoslov. Math. J.*, **10**, 258—270.

[84] Landau, L. D. and Lifshits, I. M. (1969), *A Concise Course of Theoretical Physics, 1, Mechanics, Electrodynamics*, Nauka, Moscow (in Russian).
Ландау, Л. Д. и Лифшиц, И. М. (1969), *Краткий курс теоретической физики, 1, Механика, электродинамика*, Наука, Москва.

[85] Lasota, A. (1963), Sur la distance entre les zéros de l'équation différentielle linéaire du troisième ordre, *Ann. Pol. Math.*, **13**, 129—132.

[86] Lazer, A. C. (1966), The Behavior of Solutions of the Differential Equation $y'''+p(x)y'+q(x)y=0$, *Pac. J. Math.*, **17**, 435—466.

[87] Leibenzon, Z. L. (1966), An Inverse Problem of Spectral Analysis of Ordinary Differential Operators of Higher Order, *Trudy Mosk. Mat. Obshch.*, **15**, 70—144 (in Russian).
Лайбензон, З. Л. (1966), Обратная проблема спектрального анализа обыкновенных дифференциальных операторов высшего порядка, *Труды Моск. мат. общ.*, **15**, 70—144.

[88] Leighton, W. and Nehari, Z. (1958), On the Oscillation of Solutions of Self-adjoint Linear Differential Equations of the Fourth Order, *Trans. Amer. Math. Soc.*, **89**, 325—377.

[89] Levin, A. Yu. (1963), Some Questions Concerning Oscillations of Solutions of Linear Differential Equations, *Dokl. AN SSSR*, **3**, 512—515 (in Russian).
Левин, А. Ю. (1963), Некоторые вопросы осцилляций решений линейных дифференциальных уравнений, *Докл. АН СССР*, **3**, 512—515.

[90] Ličko, J. and Švec, M. (1963), Le charactère oscillatoire des solutions de l'équation $y^{(n)}+f(x)y=0$, $n>1$, *Czechoslov. Math. J.*, **88**, 481—491.

[91] Lockschin, A. (1936), Über die Knickung eines gekrümmten Stabes, *Z. Angew. Math. Mech.*, **16**, 49—55.

[92] Matsaev, V. I. (1960), The Existence of Operator Transformations for Differential Equations of Higher Order, *Dokl. AN SSSR*, **130**, 499—502.

[93] Mammana, G. (1930), L'equazione del terzo ordine lineare omogenea. *Rc. Res. Ist. Lombardo Sci. Lett.*, **2**, 272—282.

[94] Mammana, G. (1933), Decomposizione delle equazioni differenziali lineari omogenee in prodotti di fattori simbolisi e applicazioni relative allo studio delle equazioni differenziali lineari. *Math. Z.*, **33**, 186—231.

[95] Marchenko, V. A. (1955), The Construction of the Potential Energy from the Phases of the Scattering Waves, *Dokl. AN SSSR*, **104**, 695—698.

[96] Mckelvey, Ed. (1970), *Lectures on Ordinary Differential Equations*, Academic Press, New York.

[97] Mikhlin, S. G. (1952), *Integral Equations* (in Czech, translated from Russian) — *Integrální rovnice*, Přírodovědecké vydavatelství, Praha.

[98] Myshkis, A. D. and Eventov, V. G. (1977), On a Geometrical Approach to Oscillation Properties of Third Order Linear Differential Equations, *Diff. Uravneniya*, **3**, 1047—1052 (in Russian).
Мышкис, А. Д. и Эвентов, В. Г. (1977), Об одной геометрической трактовке осцилляционных свойств линейных дифференциальных уравнений третьего порядка, *Диф. уравнения*, **3**, 1047—1052.

[99] Moravčík, J. (1960), A Note on Transformation of Solutions of Linear Differential Equations (in Slovak) — Poznámka k transformácii riešení lineárnych diferenciálnych rovníc, *Acta Fac. R. N. Univ. Comenianae, Math.*, **6**, 327—339.

[100] Moravčík, J. (1966), On a Generalization of Floquet's Theory for Ordinary n-th Order Linear Differential Equations (in Slovak) — O zobecnení Floquetovej teórie pre lineárne diferenciálne rovnice obyčajné n-tého rádu, *Čas. Pěst. Mat.*, **91**, 8—17.

[101] Moravský, L. (1967), Einige oszillatorische und asymptotische Eigenschaften der Lösungen der Differentialgleichung $y''' + p(x)y'' + 2a(x)y' + [A'(x) + x(x)]y = 0$, *Acta Fac. R. N. Univ. Comenianae, Math.*, **16**, 49—58.

[102] Moravský, L. (1974), Einige Eigenschaften der Lösungen ohne Nullstellen der linearen Differentialgleichung dritter Ordnung, *Acta Fac. R. N. Univ. Comenianae, Math.*, **29**, 69—84.

[103] Na, T. Y. (1979), *Computational Methods in Engineering Boundary Values Problems*, Academic Press, New York—London—Toronto—Sydney—San Francisco.

[104] Nehari, Z. (1957), Oscillation Criteria for Second Order Linear Differential Equations, *Trans. Amer. Math. Soc.*, **85**, 428—445.

[105] Neuman, F. (1971), Some Results on Geometrical Approach to Linear Differential Equations of the n-th Order, *Comm. Math. Univ. Carolinae*, **12**, 307—315.

[106] Neuman, F. (1972), Geometrical Approach to Linear Differential Equations of the n-th Order, *Rc. Mat.*, **5**, 579—602.

[107] Neuman, F. (1972), Oscillation in Linear Differential Equations, *Proc. Equadiff 3* (Brno), 119—125.

[108] Neuman, F. (1973), On n-dimensional Closed Curves and Periodic Solutions of Linear Differential Equations of the n-th Order, Part I, *Demonstratio Math. (Brno)*, **6**, 329—337.

[109] Neuman, F. (1974), On Two Problems about Oscillation of Linear Differential Equations of the Third Order, *J. Diff. Eqs.*, **15**, 589—596.

[110] Neuman, F. (1975), Global Transformations of Linear Differential Equations of the n-th Order, *Knižnice odb. a věd. spisů VUT Brno*, B-56, 165—171.

[111] Neumark, M. A. (1960), *Lineare Differentialoperatoren*, Akademie-Verlag, Berlin (translated from Russian).

[112] Petrov, B. N., Rutkovskii, V. Yu and Zemlyakov, S. D. (1972), *Adaptive Coordinate-parameter Control of Flying Apparatuses, Control in Cosmic Space, Vol. 1*, Nauka, Moscow (in Russian).
Петров, Б. Н., Рутковский, В. Ю. и Земляков, С. Д. (1972), *Адаптивное координатно-параметрическо управление летающими аппаратами, управление в космическом пространстве, т. 1*, Наука, Москва.

[113] Pfeiffer, G. W. (1972), Asymptotic Solutions of $y'''+qy'+ry=0$, J. *Diff. Eqs.*, **11**, 145—155.

[114] Ráb, M. (1956), Asymptotische Eigenschaften der Lösungen linearer Differentialgleichung dritter Ordnung, *Spisy přír. Fak. MU, Brno*, **374**, 1—8.

[115] Ráb, M. (1956), Asymptotic Properties of the Integrals of the Third Order Differential Linear Equation (in Czech) — Asymptotické vlastnosti integrálů diferenciální lineární rovnice 3. rádu, *Spisy Přír. Fak. MU, Brno*, **379**, 1—14.

[116] Ráb, M. (1958), On the Differential Equation $y'''+2A(x)y'+[A'(x)+w(x)]y=0$ (in Czech) — O diferenciální rovnici $y'''+2A'(x)y'+[A'(x)+w(x)]y=0$, *Mat.-Fyz. Čas.*, **8**, 115—122.

[117] Ráb, M. (1960), On a Generalization of Sansone's Theorem about Non-Oscillation of the Integral of the Differential Equation $y'''+2A(x)y'+[A'(x)-w(x)]y=0$ (in Czech) — O jistém zobecnění Sansonovy věty o neoscilaci integrálu diferenciální rovnice $y'''+2A(x)y'+[A'(x)+w(x)]y=0$, *Mat.-Fyz. Čas.*, **1**, 3—7.

[118] Regenda, J. (1976), On Some Properties of the Solutions of Linear Differential Equations of Third Order, *Math. Slov.*, **26**, 3—18.

[119] Reissig, R., Sansone, G. and Conti, R. (1969), *Nichtlineare Differentialgleichungen höherer Ordnung*, Edizione Cremonese, Roma.

[120] Rocard, V. (1949), *Dynamique générale des vibrations*, Masson et Cie, Paris.

[121] Rovder, J. (1975), Oscillation Criteria for Third Order Linear Differential Equations, *Mat. Čas.*, **25**, 231—244.

[122] Rovder, J. (1976), A Note on Comparison Theorems for Third Order Linear Differential Equations, *Math. Slov.*, **26**, 323—327.

[123] Rovder, J. (1977), Three-point Value Problem for Third Order Linear Differential Equation, *Math. Slov.*, **27**, 97—111.

[124] Sansone, G. (1948), *Equazioni differenziali nel campo reale*, Parte I, Bologna.

[125] Sansone, G. (1948), Studi sulle equazioni differenziali lineari omogenee di terzo ordine nel campo reale, *Rev. Math. Fis. Teor.*, Ser. A, 195—253.

[126] Šantavá, S. (1959), Transformation of the Integrals of a System of Two First Order Differential Equations (in Czech) — Transformace integrálů systému dvou diferenciálních rovnic 1. řádu, *Sbor. Voj. Akad. A. Zápotockého, Brno*, **8**, 3—14.

[127] Schmidt, K. (1970), A Nonlinear Boundary Value Problem, *J. Diff. Eqs.*, **7**, 527—537.

[128] Schrader, K. (1972), Second and Third Order Boundary Value Problems, *Proc. Amer. Math. Soc.*, **32**, 247—252.

[129] Šeda, V. (1965), Über die Transformation der linearen Differentialgleichungen n-ter Ordnung, I, *Čas. Pěst. Mat.*, **90**, 385—412.

[130] Šeda, V. (1967), Über die Transformation der linearen Differentialgleichungen n-ter Ordnung, II, *Čas. Pěst. Mat.*, **92**, 418—435.

[131] Šeda, V. (1967), On a Class of Linear Differential Equations of Order n, $n \geqq 3$, *Čas. Pěst. Mat.*, **92**, 247—261.

[132] Šeda, V. (1967), An Application of Green's Function in the Differential Equations, *Acta Fac. R. N. Univ. Comenianae, Math.*, **17**, 221—235.

[133] Shinn, D. (1938), Existence Theorems for the Quasi-differential Equation of the n-th Order, *Dokl. AN SSSR*, **18**, 515—518.

[134] Stewartson, K. (1949), Correlated Compressible and Incompressible Boundary Layers, *Proc. Roy. Soc.*, **200**, 84—100.

[135] Švec, M. (1957), Sur une proprieté des integrales de l'équation $y^{(n)} + Q(x)y = 0$, $n = 3, 4$, *Czechoslov. Math. J.*, **7**, 450—461.

[136] Švec, M. (1963), Asymptotische Darstellung der Lösungen der Differentialgleichung $y^{(n)} + Q(x)y = 0$, $n = 3, 4$, *Czechoslov. Math. J.*, **12**, 572—581.

[137] Švec, M. (1965), Einige asymptotische und oszillatorische Eigenschaften der Differentialgleichung $y''' + A(x)y' + B(x)y = 0$, *Czechoslov. Math. J.*, **15**, 378—393.

[138] Švec, M. (1965), Some Remarks on the Third Order Linear Differential Equation, *Czechoslov. Math. J.*, **15**, 42—49 (in Russian).
Швец, М. (1965): Несколько замечаний о линейном дифференциальном уравнений третьего порядка. *Czechoslov. Math. J.*, **15**, 1965, 42—49.

[139] Swanson, C. A. (1968), *Comparison and Oscillation Theory of Linear Differential Equations*, Academic Press, New York—London.

[140] Utz, W. R. (1970), Oscillating Solutions of Third Order Differential Equations, *Proc. Amer. Math. Soc.*, **26**, 273—276.

[141] Vencko, J. (1979), Bemerkungen zu einer Randwertaufgabe in drei Punkten für die Differentialgleichung dritter Ordnung, *Acta Fac. R. N. Univ. Comenianae, Math.*, **34**, 69—73.

[142] Vilari, G. (1958), Sul carattere oscillatorio delle soluzioni delle equazioni differenziali lineari omogenee del terzo ordine, *Boll. U. Mat. Ital.*, **3**, 73—78.

[143] Vilari, G. (1960), Contributi allo studio asintotico delle equazioni $x'''(t) + p(t)x(t) = 0$, *Ann. Mat. Pura Appl.*, **51**, 301—328.

[144] Waltman, P. J. (1966), Oscillation Criteria for Third Order Nonlinear Differential Equations, *Pac. J. Math.*, **18**, 385—389.
[145] Wintner, A. (1948), Asymptotic Integrations of the Adiabatic Oscillator in Its Hyperbolic Range, *Duke Math. J.*, **15**, 55—67.
[146] Zlámal, M. (1951), Asymptotic Properties of the Solutions of the Third Order Linear Differential Equations, *Spisy Přír. Fak. MU (Brno)*, **329**, 159—167.

Subject Index

TRANSMITTER BIOCHEMISTRY OF HUMAN BRAIN TISSUE